HYDROGEOLOGY
LABORATORY MANUAL

HYDROGEOLOGY LABORATORY MANUAL

Second Edition

KEENAN LEE
Colorado School of Mines

C.W. FETTER
University of Wisconsin—Oshkosh

JOHN E. MCCRAY
Colorado School of Mines

Pearson Education

Pearson Education, Inc.
Upper Saddle River, New Jersey 07458

Library of Congress Cataloging-in-Publication Data

Lee, Keenan,
Hydrogeology laboratory manual/Keenan Lee, C.W. Fetter, John E. McCray.
p. cm.
Includes bibliographical references.
ISBN 0-13-046549-6
1. Hydrogeology—Laboratory manuals. I. Fetter, C.W. (Charles Willard), 1942 II. McCray, John E. III. Title

GB1004.2.L44 2003
551.49'078—dc21

2002029329

Senior Editor: Patrick Lynch
Senior Marketing Manager: Christine Henry
Assistant Editor: Melanie van Benthuysen
Executive Managing Editor: Kathleen Schiaparelli
Assistant Vice President of Production and Manufacturing: David W. Riccardi
Manufacturing Manager: Trudy Pisciotti
Assistant Manufacturing Manager: Michael Bell
Manufacturing Buyer: Lynda Castillo
Production and Composition Services: Prepare Inc.
Art Studio: Prepare Inc.

Printed in the United States of America.

10 9 8 7 6 5 4 3 2 1

ISBN 0-13-046549-6

Pearson Education Ltd., London
Pearson Education Australia Pty. Ltd., Sydney
Pearson Education Singapore, Pte. Ltd.
Pearson Education North Asia Ltd., Hong Kong
Pearson Education Canada, Inc., Toronto
Pearson Educación de Mexico, S.A. de C.V.
Pearson Education—Japan, Tokyo
Pearson Education Malaysia, Pte. Ltd.
Pearson Education, Inc., Upper Saddle River, New Jersey

CONTENTS

SUPPLEMENTAL MATERIALS ON CD

Programs

PARTICLEFLOW Program	TP1.0	<folder>
TOPODRIVE Program	TP1.0	<folder>
Theis Curve-Fitting Program	THCVFIT	<folder>

Please refer to the "readme" files for instructions on the installation and use of these programs, especially when using a Mac OS.

Appendices

5 Common Phreatophytes—Western U.S.A.	APNDX5.DOC
6 Water Use, United States	APNDX6.DOC
7 Sedimentary Bedrock Aquifers of the Denver Basin	APNDX7.DOC

Data Files

Lab 5 Data	LAB5DATA.XLS
Lab 13 Data	LAB13.XLS
Lab 14 Data	LAB14.XLS
Original Lab 3 Data	OL3_DATA.XLS
$W(u)$ versus u, Theis Type Curve	W(u).XLS

Original Lab 3

Hydrology of the Upper Rio Grande Drainage Basin, Colorado	ORIGLAB3.DOC

Problems

1 Flood Frequency and Probability	PROBLEM1.DOC
2 Groundwater Chemical Evolution	PROBLEM2.DOC
3 Permeability and Hydraulic Conductivity	PROBLEM3.DOC
4 Darcy Application 1	PROBLEM4.DOC
5 Darcy Applications 2 and 3	PROBLEM5.DOC
6 Anisotropy of Multiple Isotropic Layers	PROBLEM6.DOC
7 Groundwater Refraction	PROBLEM7.DOC
8 Unconfined Flownet	PROBLEM8.DOC
9 Seismic Refraction	PROBLEM9.DOC

PREFACE

This manual is designed for use in the laboratory portion of an undergraduate course in hydrogeology or groundwater engineering. At CSM, these exercises are used in a three-hour lab that accompanies three lectures per week for one semester.

Most of the lab exercises can be used in any sequence, but they should be coordinated with lectures because they both supplement and complement lecture topics in hydrogeology. Two of the labs should be done in a laboratory that has water and sink drains, where you will be working with apparatus to measure hydraulic properties of porous media. One lab period will be spent observing an aquifer test in the field.

Several of the labs incorporate tasks that are best performed on personal computers, and the three contaminant labs are best run in a computer lab unless students have portable computers. Data files are provided on the enclosed CD that can be input to spreadsheet programs for easier calculation. The numerical model PARTICLEFLOW (used in Lab 15) requires a computer.

The final labs consist of an extended problem that should be spread over three lab periods. There are two choices of problems for this capstone mini-project: (1) The Seymour Hazardous Waste Site problem is a structured series of three labs that deals with the hydrogeology of a contaminated aquifer in the humid midwestern United States; (2) the Groundwater Basin Analysis is a quite open-ended problem, with emphasis on determining the interaction of surface water and groundwater in an arid area and predicting the future of irrigation in the basin.

Some of the exercises used in the Manual have been so modified over the years that it's difficult to recognize their origins, but parts of some of the exercises are based on material from colleagues at CSM and Stanford University. I thank the students and teaching assistants who helped test and refine these exercises. The Manual was improved considerably by helpful criticism provided by Professors Regina M. Capuano, University of Houston; David Huntley, San Diego State University; James A. Saunders, Auburn University; H. Leonard Vacher, University of South Florida; and several anonymous reviewers.

Keenan Lee

A complete hydrogeology course should include a lab as well as lecture. Until now, faculty members have been forced to develop their own custom lab manual, as no commercial manual was available. I was excited to learn of Keenan Lee's plans to publish a hydrogeology lab manual, and I was especially pleased when he invited me to contribute lab exercises from the Seymour Recycling Corporation hazardous waste site.

Hydrogeology is best learned through a hands-on approach. The material in this lab manual, in combination with the end-of-chapter problems found in textbooks, will aid students in developing a deeper appreciation for the science and art of hydrogeology.

C. W. Fetter

The physical and chemical processes that govern groundwater flow and contaminant movement in the subsurface have remained unchanged through time. Our understanding of these processes, however, and our philosophy about which processes should be incorporated into a course in subsurface hydrology, change over time. Two areas that have certainly gained much attention from academicians and practitioners alike during the last ten years are contaminant transport and computer modeling. When Keenan asked me to contribute some exercises on these two topics, I was grateful for the opportunity. I realize that only a few hydrogeology textbooks include detailed material on contaminant transport or on computer modeling, but contaminant transport theory and application commonly comprise up to half of the material in a university course in subsurface hydrology. Some departments offer two semester courses: a traditional course in hydrogeology and a follow-up course in contaminant transport.

The three new contaminant transport labs, along with the exercises for the Seymour hazardous waste site, are intended to complement today's hydrogeology courses. Each of these laboratory exercises also includes computer modeling.

I would like to thank Dr. Paul Hseih, of the U.S. Geological Survey, for providing the model PARTICLEFLOW used in Lab 15. This model is very simple to use, but it is considerably more complex behind the computer screen. Special thanks also go to Kyle Murray, a doctoral student at CSM, for designing the slug-test exercise and for critiquing the contaminant-transport exercises. Finally, Keenan Lee greatly improved the user-friendliness of these new exercises by applying his wealth of teaching experience to the original drafts.

John E. McCray

NOTE TO USERS OF THE FIRST EDITION

The manual now contains 21 lab exercises, six of which are new. The new labs include two regional aquifer studies, one in Colorado and one on the Gulf Coast, one aquifer-testing lab using slug tests, and three contaminant-transport labs dealing with use of tracers to determine aquifer and contaminant parameters and modeling retardation, biodegradation, and aquifer heterogeneity.

All material in the first edition is retained in the second edition, although some of it has been moved to the enclosed CD. One original lab exercise, all supplemental problems, and three of seven appendices have moved to the CD.

NOTE TO INSTRUCTORS

An Instructor's Manual is available at no cost from Prentice Hall.

HYDROGEOLOGY
LABORATORY MANUAL

LAB 1

WATER BUDGET OF MONO LAKE: PRECIPITATION AND EVAPORATION

PURPOSE: Familiarize you with components of the hydrologic cycle, hydrologic data sources, and techniques for analyzing these data. Become familiar with hydrologic units in metric units, which are convenient, and American (or English) units, which are necessary in the United States. This first lab will deal with precipitation and evaporation.

OBJECTIVES: Analyze hydrographic data to determine quantitative values for precipitation and evaporation.

PROBLEM: Determine the average annual groundwater flow into Mono Lake, CA.

APPROACH: Analyze the water budget for Mono Lake. Determine precipitation and evaporation.

MATERIALS: You will need a straightedge, compass (with pencil), calculator, and access to the Internet (can be done before lab). A spreadsheet program will make calculations easier.

ASSIGNMENT: Read *Hydrology of Mono Basin* below. Additional information is available in *Case Study: Mono Lake*, Fetter, 2001, pp. 9–11.[1]

Before or during lab, access the Internet to retrieve precipitation data for Table 1.1 (see *Climate* section that follows).

HYDROLOGY OF MONO BASIN

Location

Mono Basin is an intermontane, closed drainage basin in central Mono County, CA, and Mineral County, NV (Fig. 1.1). The basin is about 300 kilometers east of San Francisco and forms part of the eastern boundary of Yosemite National Park. Lee Vining and June Lake, CA, are the only two towns within the basin.

Basin Morphometry

The shape of the Mono Basin is slightly elongate northeast–southwest, with dimensions of about 50 km by 30 km (30 mi by 20 mi). The enclosed area is 1748 km^2 (675 mi^2), including Mono Lake (215 km^2, 83 mi^2). The lake is fairly elliptical, about 22 km (13 mi) east–west by about 16 km (10 mi) north–south (Fig. 1.1).

The basin floor is relatively flat, sloping gently upward from Mono Lake at 1948 m (6390 ft) to the base of the surrounding rim of mountains at about 2200 m (7200 ft) (Fig. 1.2). The Bodie Hills to the north rise fairly steeply to elevations of about 2500 m (8200 ft), and in the south, the narrow arcuate chain of the Mono Craters, about 2700 m (9000 ft), extend northward to within 1.5 km of the lake.

West of Mono Lake, the Sierra Nevadas rise abruptly from the lake and culminate in snowy crests at elevations near 4000 m (Mt. Lyell, 3997 m, 13,114 ft; Mt. Dana, 3979 m, 13,053 ft; Mt. Gibbs, 3890 m, 12,764 ft). In this region, the Sierra Nevada Divide is the western drainage boundary of the Mono Basin. The mountains exhibit rugged relief

[1]Some references are made to C.W. Fetter, 2001, *Applied Hydrogeology*, 4th edition: Prentice-Hall, Upper Saddle River, NJ.

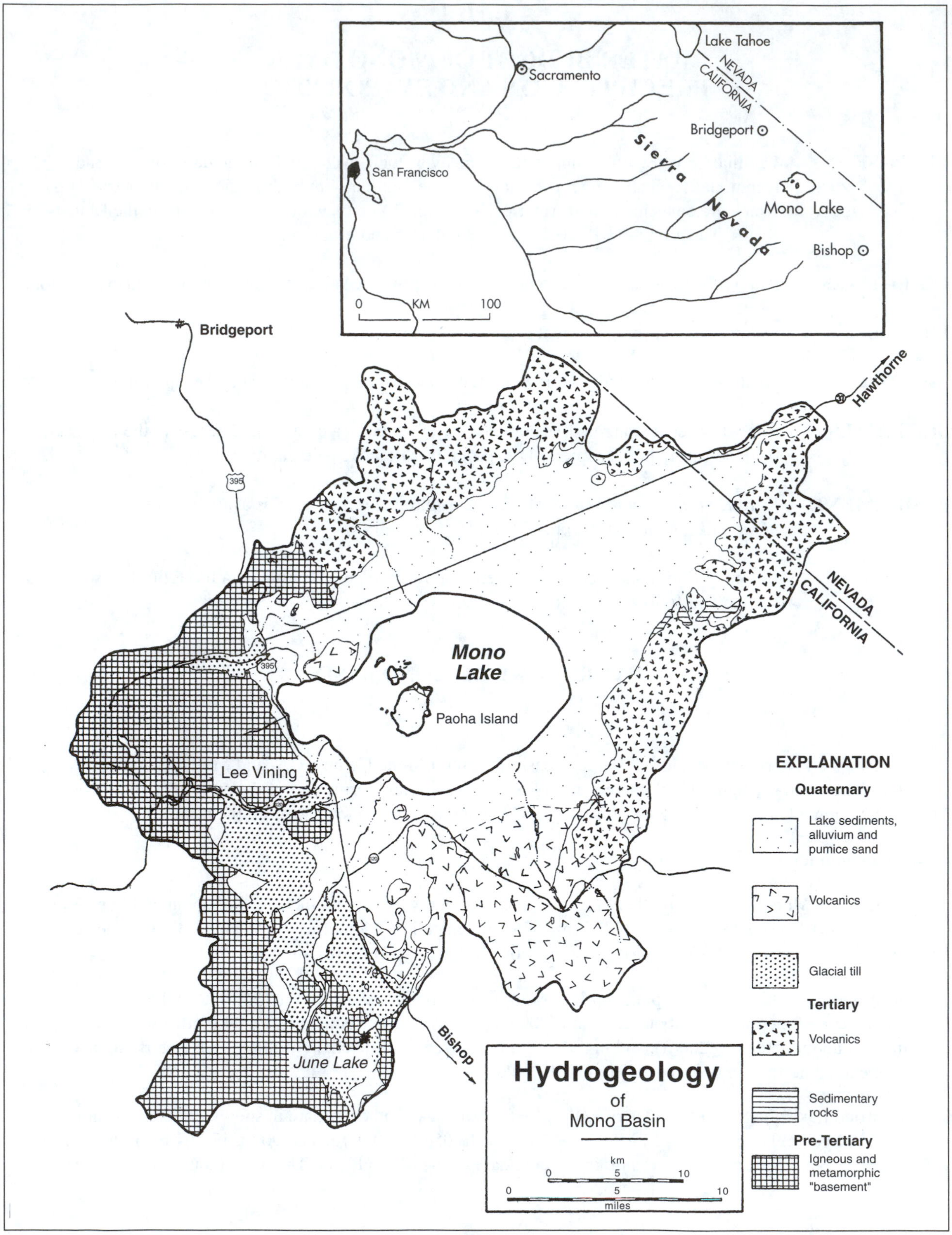

Figure 1.1—Index map and hydrogeologic map of the Mono Basin (Lee, 1969, Figure 6).

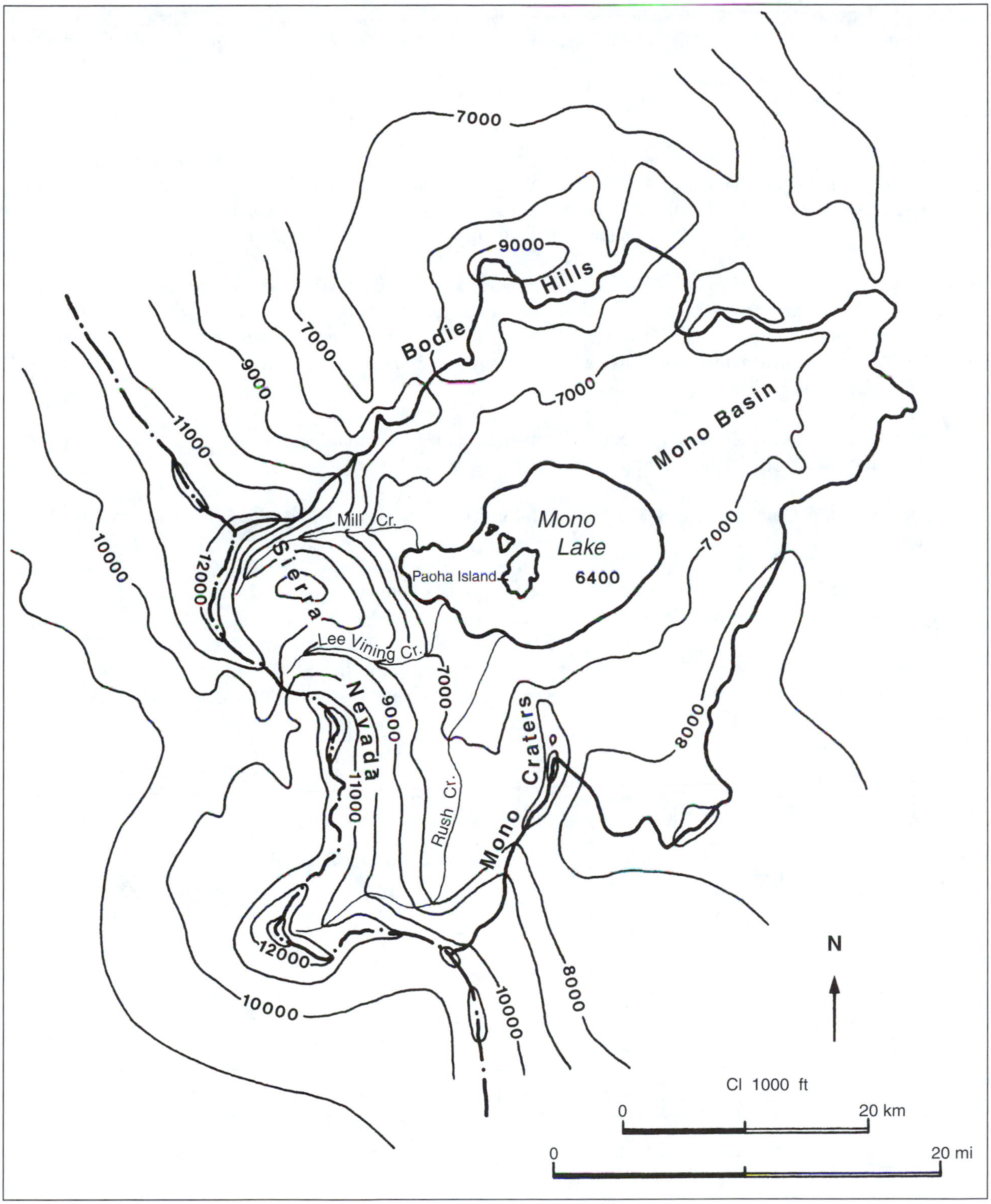

Figure 1.2—Topographic map of the Mono Lake–Sierra Nevada region (contour interval 1000 ft; heavy line is the divide of Mono Basin; dash–dot line is the divide of the Sierra Nevada).

characteristic of glacially sculpted mountains, with deep, narrow, U-shaped valleys and horn-shaped peaks. The steep eastern flank of the Sierras is accentuated by a very steep scarp (500 m/km, 2600 ft/mi) immediately alongside Mono Lake.

Climate

The climate of the Mono Basin is continental, with cold winters, during which most of the annual precipitation occurs, and dry summers with hot days and cool nights. The basin lies in an area of strong gradients; elevations rise precipitously from the basin floor to the crest of the Sierra Nevadas, and as a consequence, mean annual temperatures drop with increasing elevation (environmental lapse rate is −5.8°C/1000 m).

The United States Weather Bureau (USWB, now Environmental Data Services of NOAA) records meteorological observations at three stations within the Mono Basin and at several stations nearby. In addition, the Los Angeles Department of Water and Power maintains a station at Cain Ranch. These stations are shown on the map in Figure 1.3. Pertinent data are summarized in Table 1.1, *except for Ellery Lake, for which you will need to retrieve precipitation data from the Internet.*

1. Go to the California Data Exchange Center, maintained by the California Department of Water Resources as the access point to hydrologic data, at: http://cdec.water.ca.gov.

2. From the "CDEC Quick Search," select "Station Information" to find the 3-letter code for the Ellery Lake station.

3. From the "CDEC Quick Search," select "Download CSV Data," use the "Monthly Data Sorted By Years" form, and retrieve data from the entire period of record (sensor number is 2).

4. Drop these data into a spreadsheet for quick calculation of average precipitation. Calculate average monthly precipitation for the period of record, and sum these values to obtain the mean annual precipitation for the period of record.

Table 1.1—PRECIPITATION, MONO BASIN AREA

	Elevation		Average Annual Precipitation		
Station	**m**	**ft**	**cm**	**in.**	**from records of**
Bodie	2551	8370	41.1	16.2	1965–1968, 1996–1999 (8)
Benton	1661	5450	19.3	7.6	1965–1968 (4)
Ellery Lake	2940	9645			1924–2000 (75)
Gem Lake	2760	9054	53.3	21.0	1924–2000 (75)
Mark Twain Camp	2204	7230	17.3	6.8	1950–1955 (4)
Mono Lake	1966	6450	34.3	13.5	1951–1980, 1982–1988 (36)
Cain Ranch	2097	6880	28.1	11.1	1921–1964 (34)

WATER BUDGET

Write a continuity equation for the water budget of Mono Lake, including all components that you think might possibly be significant (the continuity equation is also known as the law of mass conservation and is referred to as the *hydrologic equation* by Fetter). Consider addition to the lake as positive and removal as negative.

PRECIPITATION

Arithmetic Average Method

Precipitation is measured at recording stations that provide point data of linear depth. In order to determine the volume of precipitation falling within a basin, these point data somehow must be extrapolated over an area that they represent (the *effective uniform depth* of Fetter). The easiest way of doing this is the arithmetic averaging method, in which one simply multiplies the average of the point data by the area of the entire basin.

1. Determine the average annual precipitation (m^3) by averaging the precipitation data for stations within the Mono Basin (Table 1.1). Calculate precipitation in the Mono Basin and, separately, on Mono Lake.

Thiessen Method

A more accurate method for determining basin precipitation, the Thiessen method, accounts for nonuniform distribution of recording stations by weighting each data point differently, according to the percentage of the basin the station represents. The method assumes that the precipitation at any point in the basin is that of the nearest station, or (another way of saying this) the precipitation is assumed to vary linearly between stations. A description of the Thiessen method is available in most textbooks (e.g., Fetter, 2001, pp. 34–37).

2. Determine the average annual precipitation on Mono Lake (m^3) by the Thiessen method (do this for the *lake only*), using the map in Figure 1.3. Areas within each polygon normally are measured with a planimeter, but for the sake of this exercise, you can estimate the areas by "counting squares," using quad-ruled paper.

Compare your result with precipitation calculated by the averaging method (No. 1).

Isohyetal Method

Where orographic effects are significant, the isohyetal method gives a more accurate estimate because it accounts for topography. The method takes into account not only a nonuniform distribution of stations, but it allows for nonlinear variations in precipitation as well, as might be expected along a mountain range.

3. Plot precipitation as a function of elevation on the graph. Does there appear to be an orographic effect?

   Estimate a regression line, and determine the precipitation gradient.

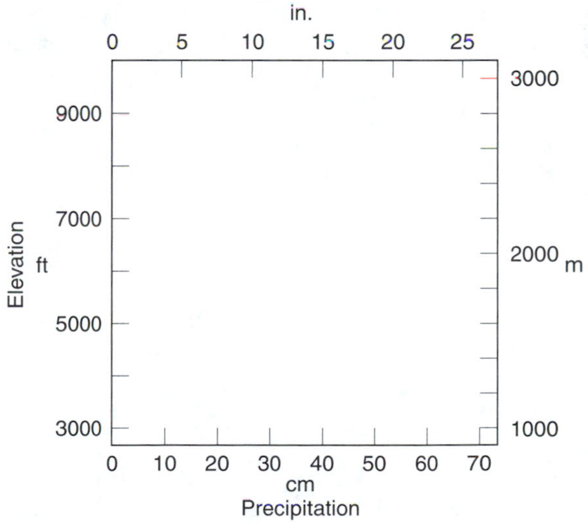

Here in the rain shadow of the Sierras, where gradients are strong, the isohyetal method gives the most effective estimate of precipitation.

Using the isohyetal method, precipitation values are plotted at each station and contoured with lines of equal precipitation, or isohyets, taking into account the topography and using your knowledge of prevailing wind directions and orographic precipitation. After the isohyetal map is completed, precipitation is determined by measuring the areas between successive isohyets and multiplying each area by the average precipitation between its bounding contours.

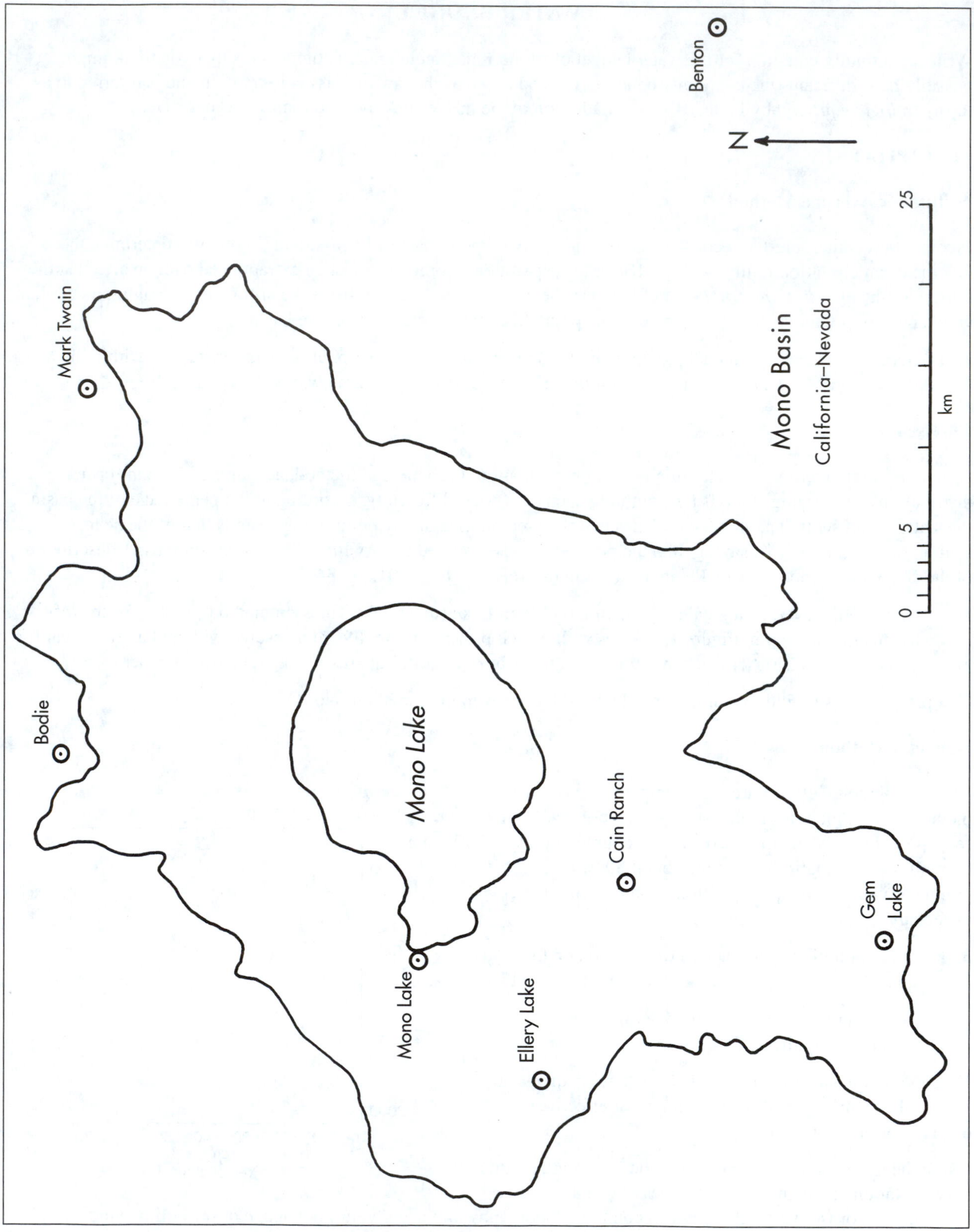

Figure 1.3—Precipitation stations in and around Mono Basin.

4. Construct an isohyetal map of the Mono Basin on Figure 1.4, making reference to the topographic map in Figure 1.2. Compute the average annual precipitation on Mono Lake by this method. Compare your result with precipitation determined by the other two methods (Nos. 1 and 2).

EVAPORATION

You will estimate the evaporation at Mono Lake in two ways: (1) by referring to a published map and (2) by reducing data from an evaporation pan at Cain Ranch.

1. Estimate average annual evaporation, or lake evaporation (m), by viewing the evaporation map in Figure 1.5. For conversion tables from American units to SI units, refer to your textbook (e.g., Fetter, 2001, Appendices 7–9).

The data for the lake evaporation map compiled by Kohler and others (1959) were derived from measurements of water evaporated from a standardized pan: a 4-ft-diameter, galvanized pan known as the Class A land pan. These measurements of pan evaporation are shown on the map in Figure 1.6.

Because evaporation pans have greater evaporation rates than lakes, typically about 140 percent, they must be corrected by applying an empirically derived pan coefficient. The variation in pan coefficients is shown on a similar map in Figure 1.7. In effect, the lake evaporation map you used (Fig. 1.5) is a derivative of the other two maps—measured pan evaporation (Fig. 1.6) multiplied by a pan coefficient (Fig. 1.7).

2. The following pan data were recorded at Cain Ranch.

   Cain Ranch Station

   Elevation: 6880 ft

   Class A evaporation pan, water level held constant by float valve; volume of water needed to recharge pan (corrected for precipitation) is recorded.

May	33.71 gal	August	73.94 gal
June	51.41 gal	September	49.41 gal
July	94.10 gal	October	33.51 gal

   (a) Calculate the pan evaporation (m) for this period.

   (b) Data for the winter months are usually not recorded because in much of the United States, freezing prevents measurement. Extrapolate for annual pan evaporation using the map in Figure 1.8.

   (c) Correct for the pan effect, using the map in Figure 1.7 to determine annual lake evaporation.

   (d) How does this measurement compare with the reading you took directly from the map (No. 1)?

3. Calculate the annual evaporation (m^3) from Mono Lake.

LAB REPORT

Prepare a lab report summarizing the analysis to date, showing all calculations.

For all quantitative values, ensure that you report significant figures (significant digits) only (e.g., see Fetter, 2001, p. 19).

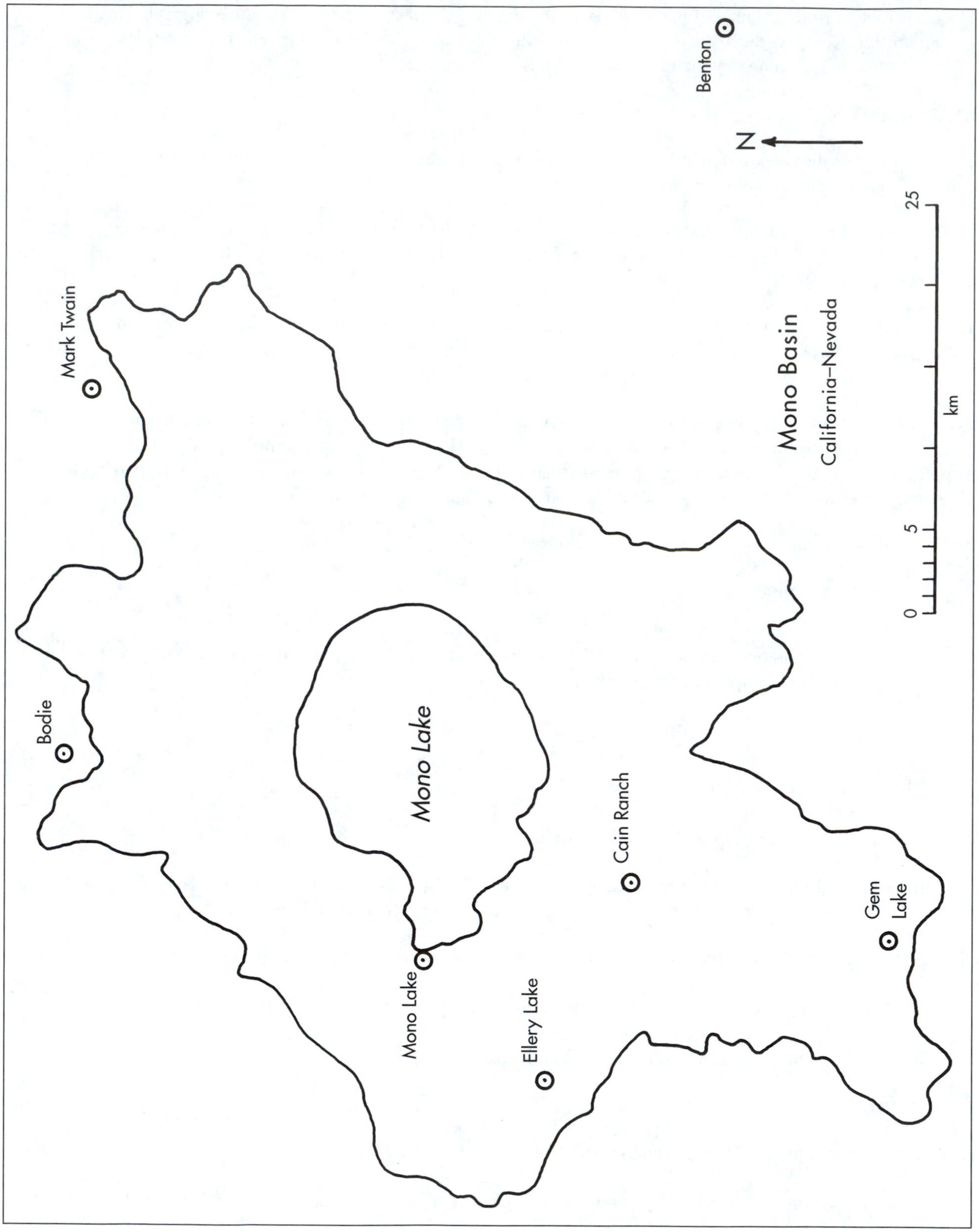

Figure 1.4—Precipitation stations in and around Mono Basin.

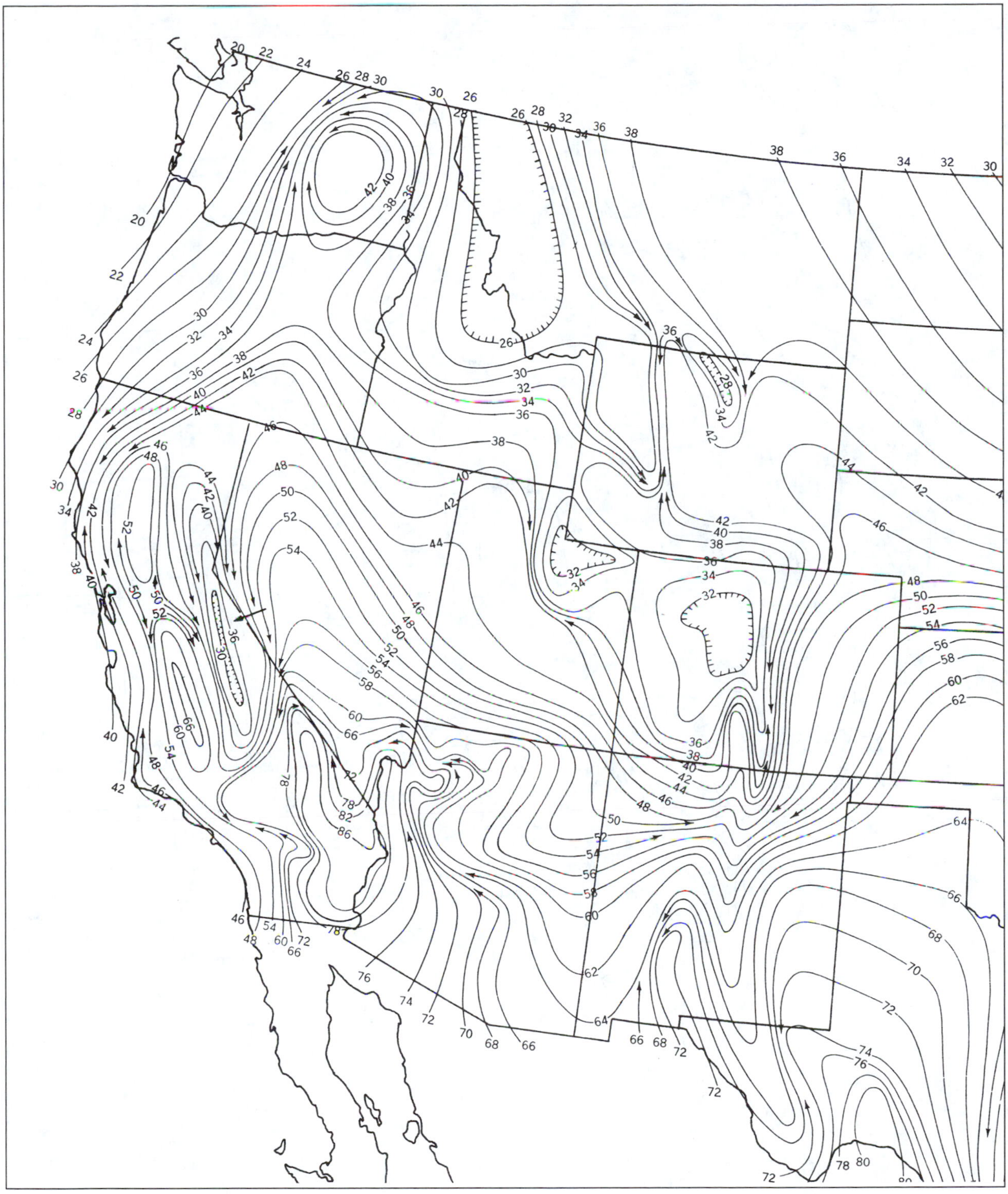

Figure 1.5—Average annual lake evaporation (in inches) in the western United States for the period 1946–1955 (from Kohler et al., 1959, Plate 2).

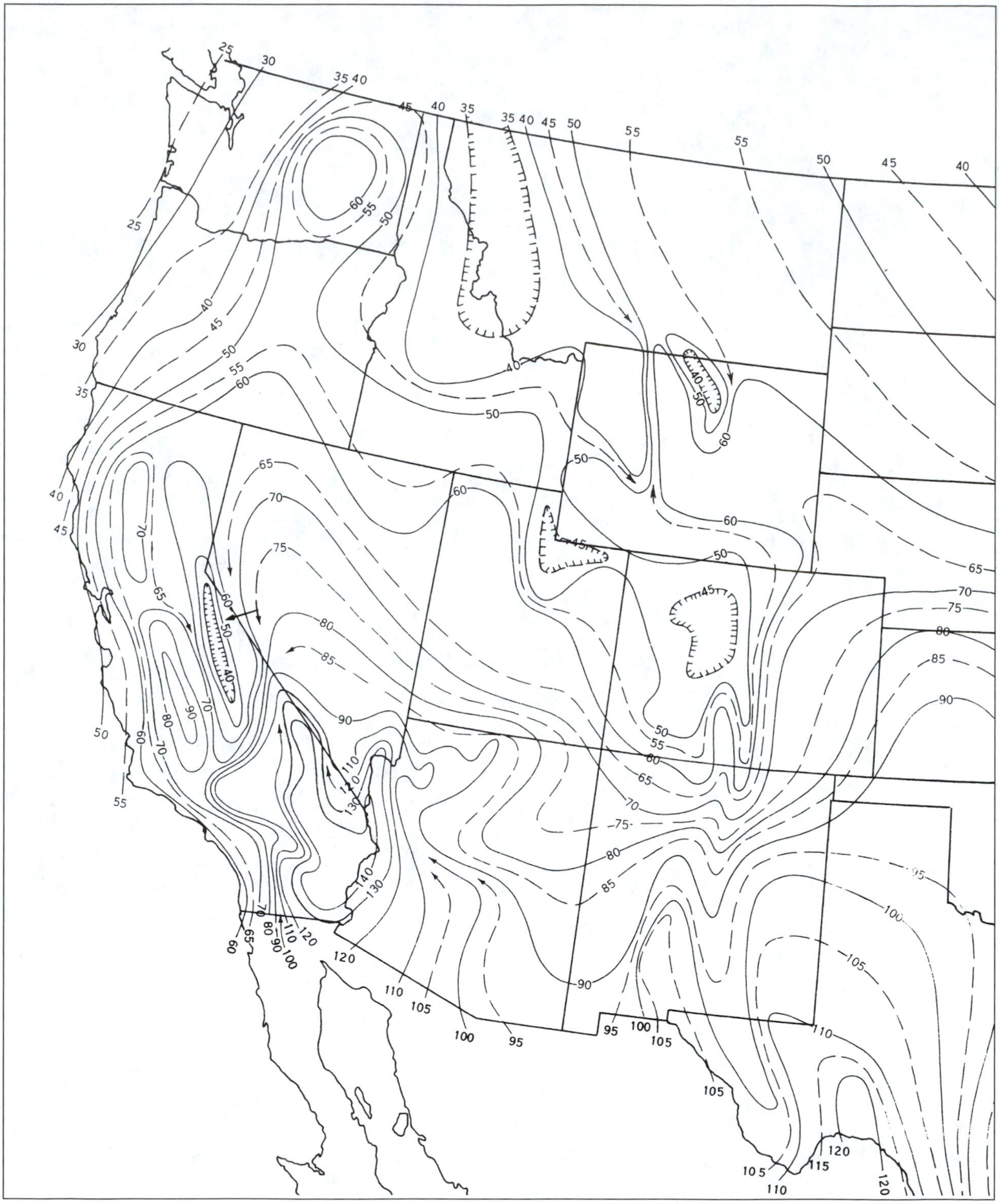

Figure 1.6—Average annual Class A pan evaporation (in inches) in the western United States (from Kohler et al., 1959, Plate 1).

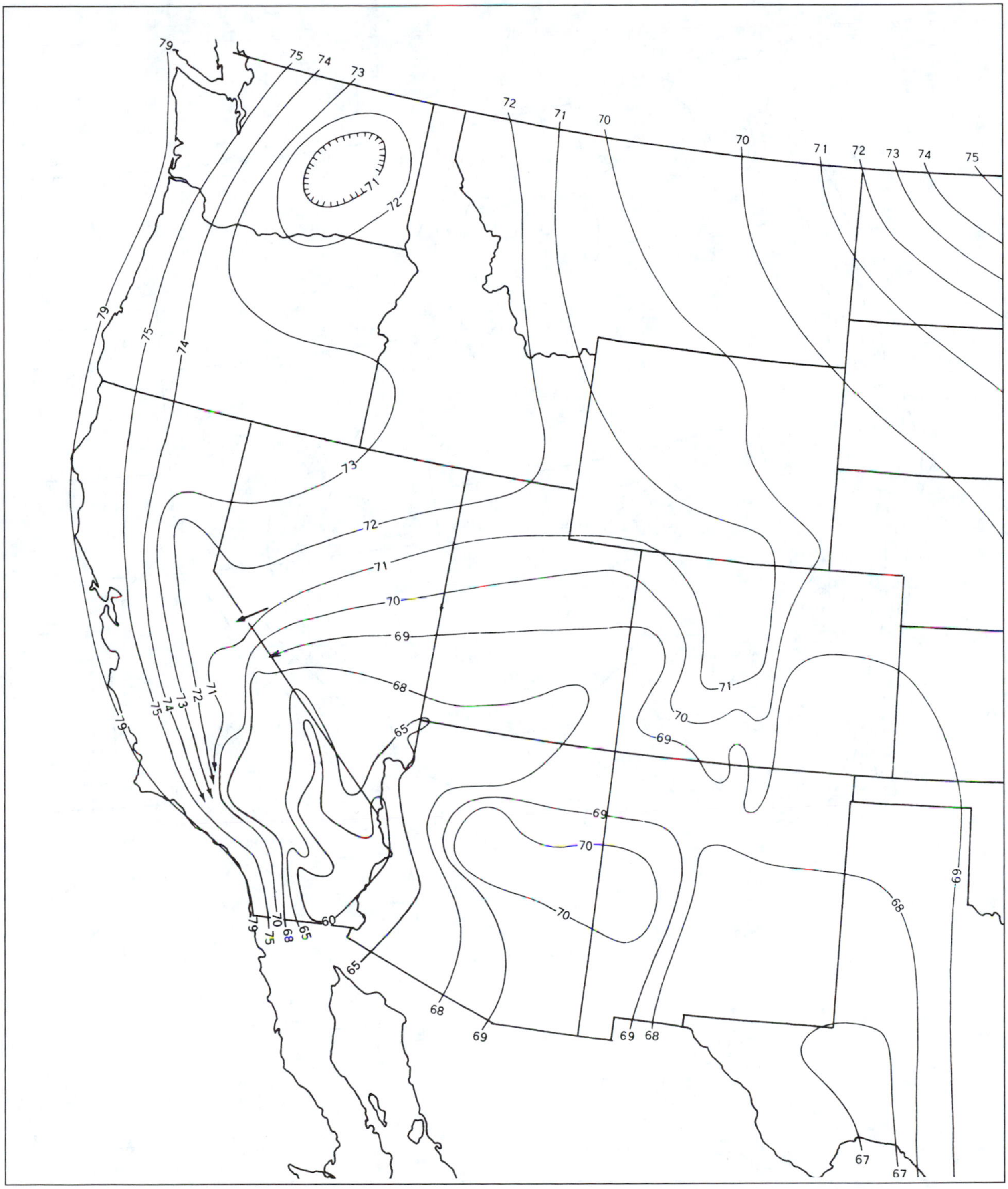

Figure 1.7—Average annual Class A pan coefficient (in percent) in the western United States (from Kohler et al., 1959, Plate 3).

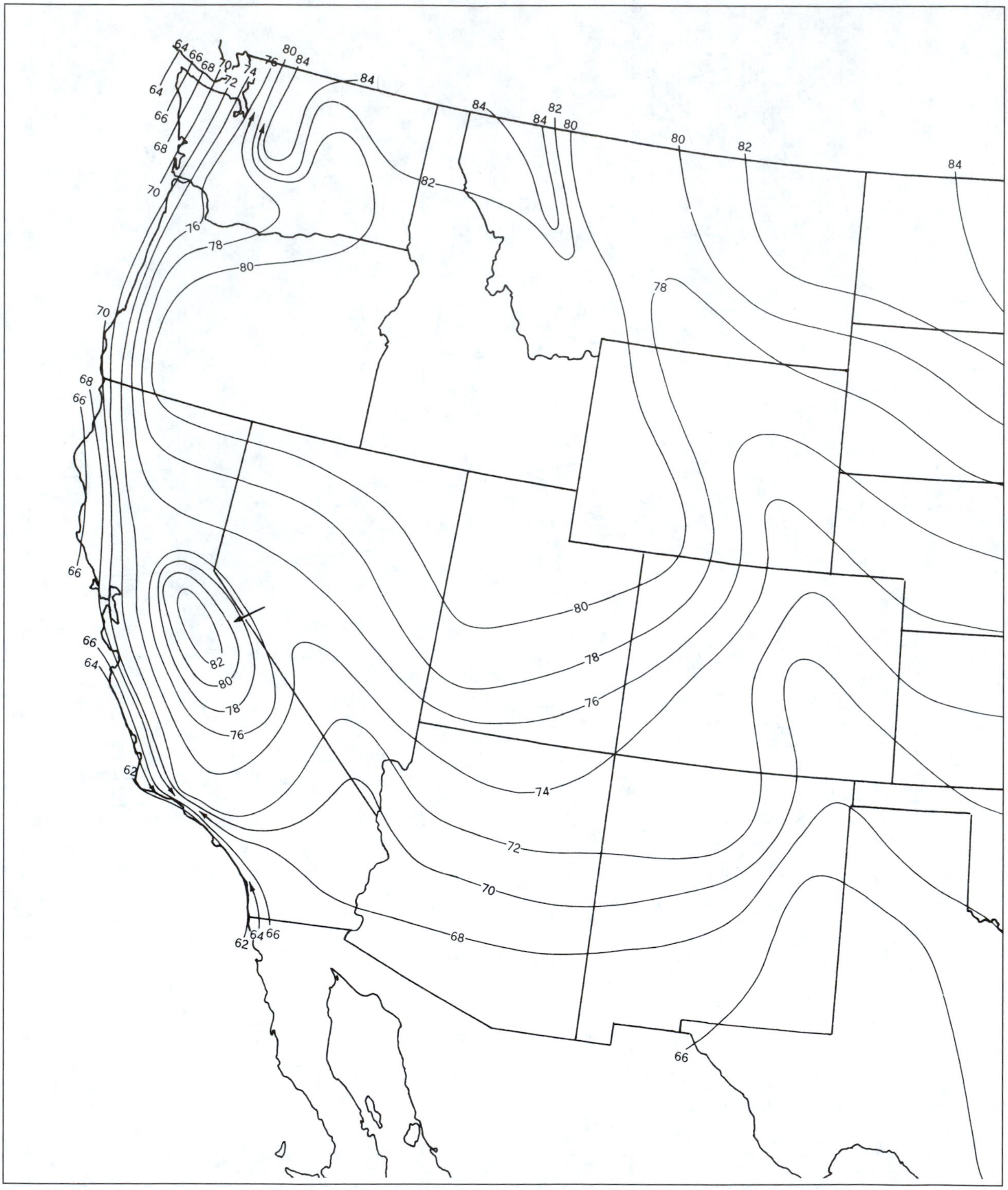

Figure 1.8—Average May–October evaporation (percent of annual) in the western United States (from Kohler et al., 1959, Plate 4.)

LAB 2

WATER BUDGET OF MONO LAKE: RUNOFF, STORAGE, AND GROUNDWATER FLOW

PURPOSE: Familiarize you with components of the hydrologic cycle, data sources, and techniques for data analysis.

OBJECTIVES: Determine runoff, change in storage, and solution of the continuity equation to determine groundwater flow into Mono Lake.

PROBLEM: Determine the average groundwater flow into Mono Lake, CA.

APPROACH: Write the general continuity equation for the water budget of Mono Lake, indicating all possible inputs and outputs.

RUNOFF

Determine the average annual runoff (m^3) from all streams flowing into Mono Lake.

1. Three large streams flow into Mono Lake, all from the Sierra Nevada (Fig. 1.2). These streams are gaged by the Los Angeles Department of Water and Power (DWP), and have the following average annual runoff.

Lee Vining Creek	87.3 cfs (cubic feet per second)
Rush Creek	44.8 cfs
Mill Creek	37.0 cfs

   What is the combined average annual runoff of these three streams, in m^3?

2. The Los Angeles DWP has been diverting water from Lee Vining, Rush, and Mill Creeks (below the gaging stations) for many years, running the diverted water through the Owens Valley aqueduct to Los Angeles. These diversions average 105, 305 acre-ft per year.

   Taking into account these diversions, what is the average annual flow into Mono Lake from these three streams, in m^3?

3. Two small streams, not gaged, were measured with simple weirs to determine their runoff.

   The smallest stream, Andy Thompson Creek, was measured with a 90° triangular-notch weir. This type of weir is most useful for discharges of less than 1 cfs. The formula for a weir with this geometry is

   $$Q = 2.54\, h^{5/2}$$

   where Q is in cfs and h is in ft.

   The measurement, taken after flow is stabilized, showed the water surface 6.9 inches above the notch. Determine runoff (m^3/yr).

Triangular-notch Weir (90°)

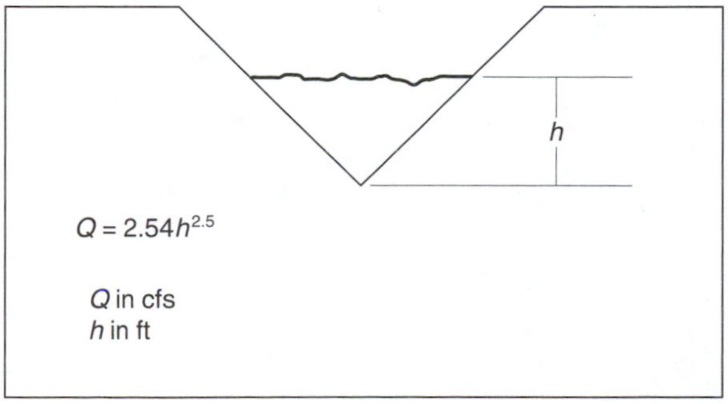

4. The larger stream required a larger weir, so a 3-foot Cipolletti weir was used. This type of weir is useful for flows of a few cfs. The Cipolletti formula is

$$Q = 3.33(L - 0.2h)h^{3/2}$$

where Q is in cfs, and L and h are in ft.

The flow stabilized at a height of 3.5 inches. Calculate discharge (m^3/yr).

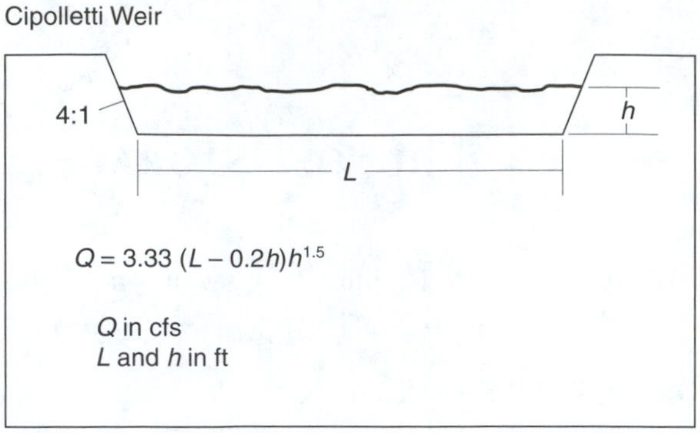

5. Numerous springs discharge along the shoreline of Mono Lake. All significant springs were monitored by weirs, and their average total flow is 5530 gpm (gallons per minute), which should be considered surface flow, or runoff.

Convert spring runoff to m^3/yr.

6. Summarize the average annual runoff into Mono Lake.

STORAGE

1. Mono Lake is more or less elliptical in map view, but morphometric studies of the lake have determined that its volume can be approximated well by the volume of a cone, as though the lake were circular with a uniformly sloping bottom. Thus, any change in storage can be calculated simply by determining the volume of the frustrum of a cone, as shown in the diagram.

From the data recorded by the Los Angeles DWP, Mono Lake stood at 6406.9 ft in June 1954, and it had fallen to 6391.2 ft by June 1964. Surface area of the lake in June 1954 was 89.4 mi^2, while the area in June 1964 was 77.0 mi^2.

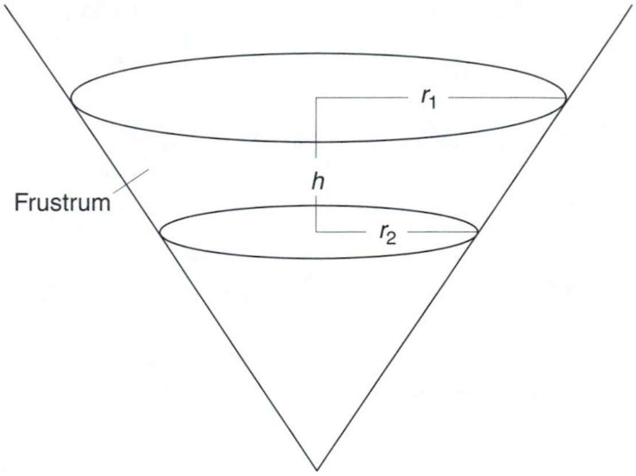

$$\text{Volume} = \frac{1}{3}\pi h\,(r_1^2 + r_1 r_2 + r_2^2)$$

Assuming that the lake is about circular, determine the respective radii of the lake in 1954 and in 1964.

Calculate the change in storage of Mono Lake over the ten-year period from June 1954 to June 1964.

2. What was the average annual change in storage over this period (m^3)?

GROUNDWATER FLOW

1. Using all the data presented so far, solve the continuity equation for average annual groundwater flow. Base this value on data from the period from June 1954 to June 1964. For precipitation, use the value of 40,000,000 m^3/yr (determined from an isohyetal analysis), and for evaporation, use the value 215,000,000 m^3/yr.

2. To arrive at this estimate, what assumptions have you made about any other elements of the water budget? Refer back to the continuity equation you wrote at the beginning of the lab.

WATER BUDGET

1. Summarize the annual water budget for Mono Lake, using the table that follows. Take care to report only significant figures.

ANNUAL WATER BUDGET FOR MONO LAKE (1954–1964)			
Inflow (m^3/yr)		**Outflow + Change in Storage (m^3/yr)**	
Runoff		Evaporation	
Precipitation			
Groundwater Flow		Change in Storage	
Total Inflow		**Total Outflow + Change in Storage**	

2. Discuss briefly the possible sources of error in the annual water budget.

3. Which element of the water budget probably contains the largest error? Why?

LAB 3

REGIONAL AQUIFER STUDY: COLORADO

PURPOSE: Introduce you to the scope and magnitude of regional aquifers and their systems.
Introduce you to groundwater occurrences in diverse aquifer materials under varying conditions.
Introduce you to the resources of the *Ground Water Atlas of the United States* and to the RASA program.

OBJECTIVES: Determine the regional aquifers in a given study area, and learn about their basic properties.
Investigate one aquifer in substantial detail, sufficient to site a municipal well.

EXERCISE OVERVIEW: You will read brief summaries of each of the regional aquifers and outline their characteristics using the *Ground Water Atlas*. You will then study one aquifer in detail to provide the background information used in selecting an aquifer for a municipal well, using the atlas and one of the RASA studies published by the U.S. Geological Survey.

Lab 4 deals with a study area along the Gulf Coast. Your instructor may ask you to study this area also, or, in fact, any other area in the world where sufficient information on regional aquifers has been published. The procedures and questions used in this exercise can be used as a template.

MATERIALS REQUIRED: *Ground Water Atlas of the United States, Segment 2 (Arizona, Colorado, New Mexico, Utha)* (Robson and Banta, 1995)
Bedrock Aquifers of the Denver Basin, Colorado—A Quantitative Water-Resources Appraisal (Robson, 1987)
Optional: Alluvial and Bedrock Aquifers of the Denver Basin (Robson, 1989)
Optional: Ground Water Resources of the Bedrock Aquifers of the Denver Basin (Romero, 1976)

STUDY AREA: Colorado contains five regional aquifers (Fig. 3.1). Western Colorado contains the Colorado Plateau regional aquifers, not considered in this study. Underlying eastern Colorado is the Central Midwest aquifer system, important in states to the east, but, although the rocks crop out along the Front Range, they are too deep to be used in Colorado, and this system also will not be studied.

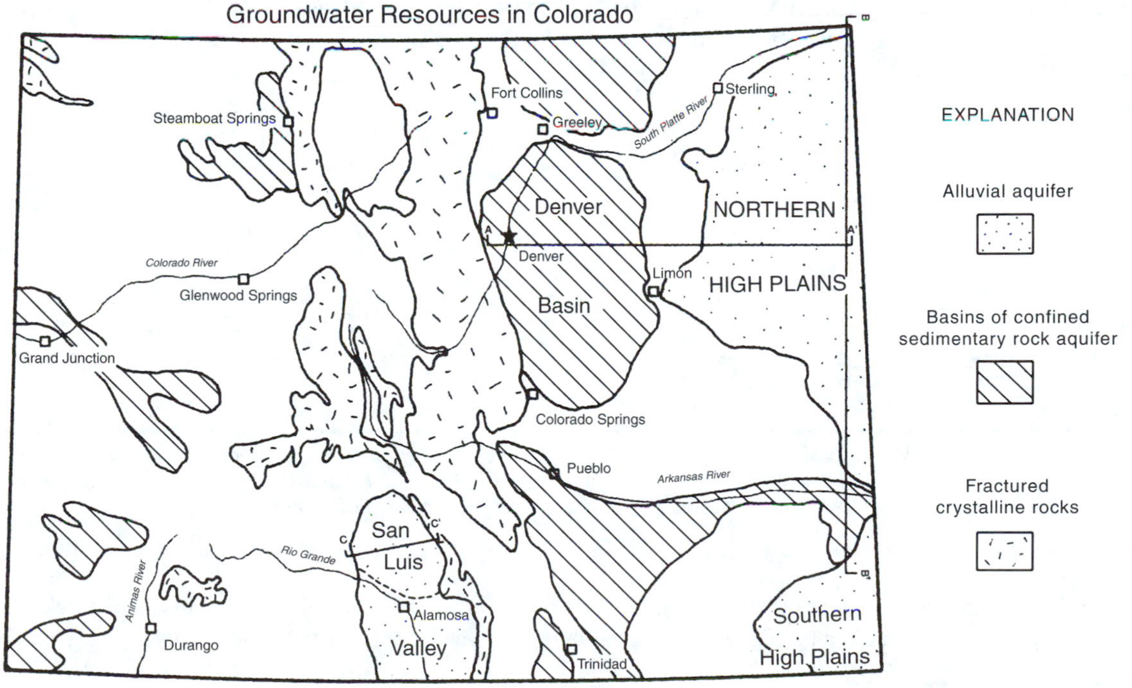

Figure 3.1—Generalized distribution of groundwater in Colorado.

This study area is central and eastern Colorado, which contains three regional aquifer systems: the sedimentary rock aquifers of the Denver basin, the unconsolidated sediments of the high and dry High Plains, and the very thick, rift-valley-fill sediments of the Rio Grande aquifer system in the San Luis Valley.

DENVER BASIN AQUIFER SYSTEM[1]

Large volumes of groundwater are contained in alluvial and bedrock aquifers in the semiarid Denver basin of eastern Colorado. Bedrock aquifers, for example, contain 1.2 times as much water as Lake Erie of the Great Lakes, yet they supply only about 9 percent of the groundwater used in the basin. Although this seems to indicate underutilization of this valuable water supply, this is not necessarily the case because many factors other than the volume of water affect the use of an aquifer. Such factors as climatic conditions, precipitation, runoff, geology and water-yielding character of the aquifers, water-level conditions, volume of recharge and discharge, legal and economic constraints, and water-quality conditions can ultimately affect the decision to use groundwater. Knowledge of the function and interaction of the various parts of this hydrologic system is important to the proper management and use of the groundwater resources of the region.

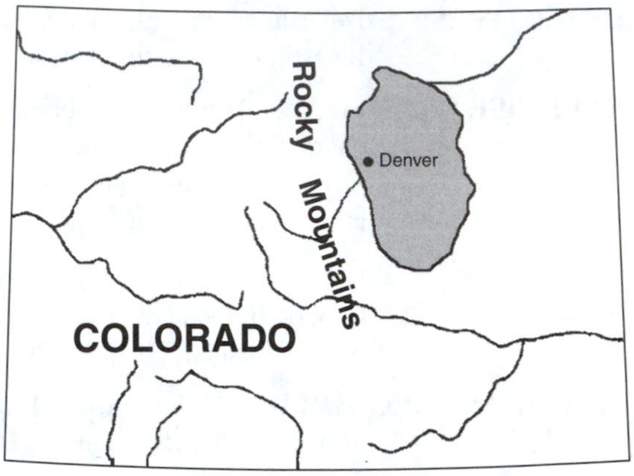

Alluvial aquifers are recharged easily from runoff and readily store and transmit the water because they consist of relatively thin deposits of gravel, sand, and clay in the valleys of principal streams. The bedrock aquifers are recharged less easily because of their greater thickness (as much as 3000 feet) and prevalent layers of shale, which retard the downward movement of water in the formations.

Although the bedrock aquifers contain more than 50 times as much water in storage as the alluvial aquifers, they do not store and transmit water as readily. For example, about 91 percent of the water pumped from wells is obtained from the alluvial aquifers, yet water-level declines generally have not exceeded 40 feet. By contrast, only 9 percent of the water pumped from wells is obtained from bedrock aquifers, yet water-level declines in these aquifers have exceeded 500 feet in some areas.

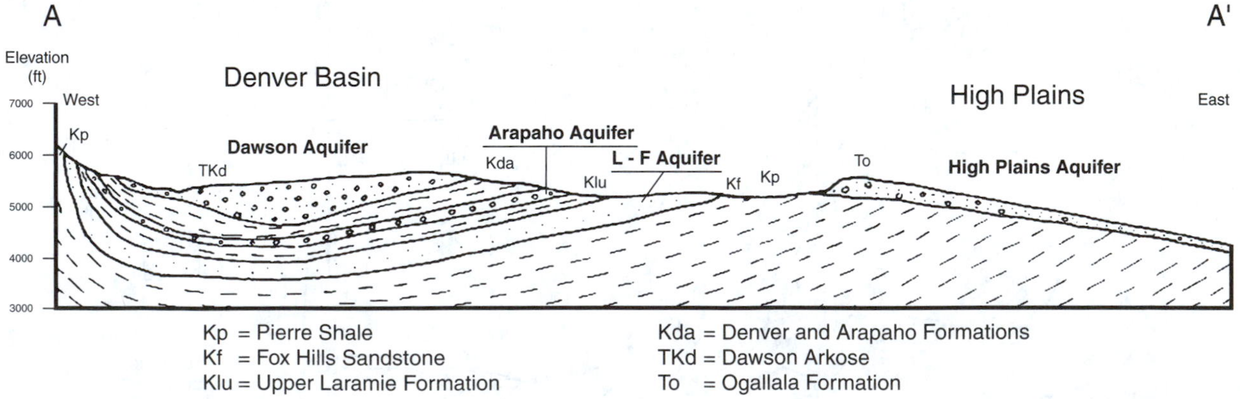

Figure 3.2—Simplified east–west cross section from Front Range across the Denver Basin and the High Plains.

Depth to the water in the alluvial aquifers generally is less than 40 feet, while depth to water in the bedrock aquifers may exceed 1000 feet in some areas (Fig. 3.2). Costs of pumping water to the surface and of maintaining existing

[1]Abstracted from Robson, 1989.

supplies in areas of rapidly declining water levels in the bedrock aquifers affect water use. Water use is also affected by the generally poorer quality water found in the alluvial aquifers and, to a lesser extent, by the greater susceptibility of the alluvial aquifers to pollution from surface sources.

Because of these factors, the alluvial aquifers are used primarily as a source of irrigation supply, which is the largest water use in the area. The bedrock aquifers are used primarily for domestic or municipal supply, which is the smaller of the two principal uses, even though the bedrock aquifers contain 50 times more stored groundwater than the alluvial aquifers.

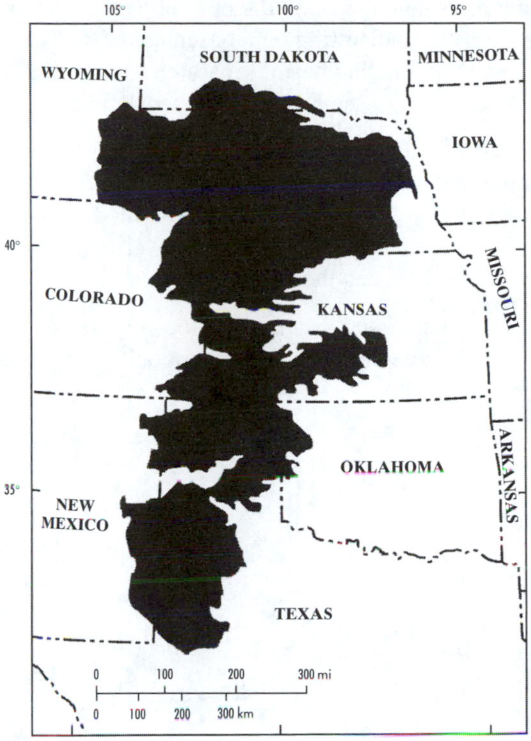

HIGH PLAINS AQUIFER[2]

The High Plains regional aquifer covers an area of 174,000 mi^2 of flat to gently rolling terrain in parts of eight states ranging from South Dakota to Texas. The High Plains regional aquifer is unconfined and consists mainly of near-surface sand and gravel deposits. The Ogallala Formation of Tertiary age, which underlies about 80 percent of the High Plains, is the principal aquifer, which consists of sediments of late Tertiary or Quaternary age, underlain by lower Tertiary to Cretaceous sedimentary rocks of low permeability named the Great Plains confining system (Fig. 3.3). The sediments that make up the aquifer are mainly alluvial, dune-sand, and valley-fill deposits.

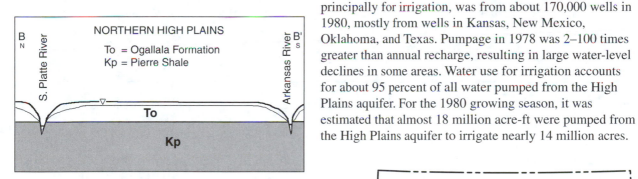

Hydraulic conductivity of the aquifer ranges from less than 25 to 300 ft/d, averaging about 60 ft/d. Specific yield ranges from less than 10 to 30 percent and averages about 15 percent. Pumpage from the High Plains aquifer, principally for irrigation, was from about 170,000 wells in 1980, mostly from wells in Kansas, New Mexico, Oklahoma, and Texas. Pumpage in 1978 was 2–100 times greater than annual recharge, resulting in large water-level declines in some areas. Water use for irrigation accounts for about 95 percent of all water pumped from the High Plains aquifer. For the 1980 growing season, it was estimated that almost 18 million acre-ft were pumped from the High Plains aquifer to irrigate nearly 14 million acres.

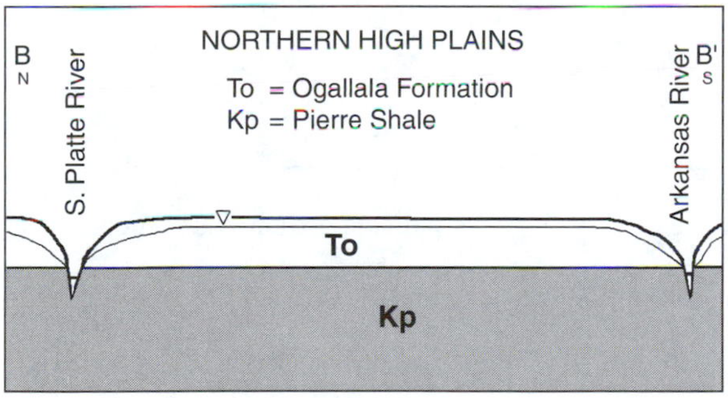

Figure 3.3—Simplified north–south cross section across the High Plains (Ogallala) Aquifer.

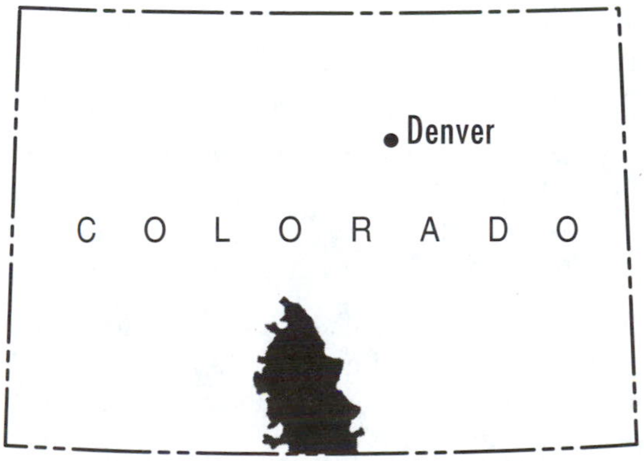

RIO GRANDE AQUIFER SYSTEM[3]

The Rio Grande aquifer system consists of thick unconsolidated sediments that fill the Neogene Rio Grande rift, extending from the San Luis Valley in Colorado through New Mexico and into Texas. The San Luis Valley is a large north-trending structural depression that is downfaulted on the eastern border and hinged on the western side. The valley

[2]From Sun and Johnson, 1994.

[3]Modified from Emery, 1971.

contains as much as 30,000 feet of alluvium, volcanic debris, and interbedded volcanic flows and tuffs of Eocene to Holocene age, all sitting on a basement of Precambrian crystalline rocks. The northern end of the San Luis Valley is a closed topographic basin, separated from the rest of the valley to the south by a low levee-like ridge along the northeast bank of the Rio Grande, shown in Figure 3.1 by a dashed line.

At least 2 billion acre-feet of water are stored in the upper 6000 feet in the San Luis Valley, in an upper unconfined aquifer and a deeper confined aquifer (Fig. 3.4). These aquifers are separated by a confining *blue clay series*, but these confining beds are discontinuous and lenticular, so it is difficult to differentiate between unconfined and confined aquifers except locally. This discontinuity in the blue clay series creates varying degrees of hydraulic connection between the aquifers.

Shallow unconfined groundwater occurs almost everywhere in the valley and extends 50–200 feet beneath the land surface. The depth to water in about half of the valley is less than 12 feet.

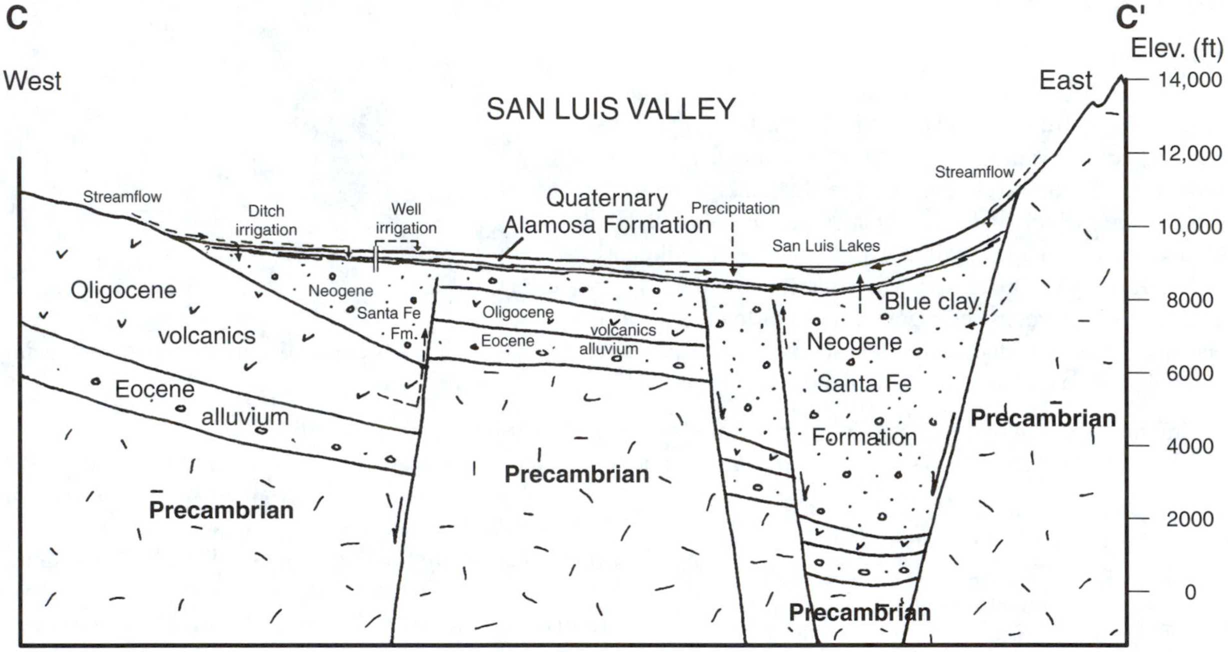

Figure 3.4—Simplified east–west cross section across the San Luis Valley. Arrows indicate groundwater recharge and flow.

Most recharge to the unconfined aquifer comes from infiltration of applied irrigation water and leakage from canals and ditches. Non-anthropogenic recharge consists of water infiltrating from the many streams entering the valley, precipitation on the valley floor, and upward leakage from the deeper confined aquifer. Discharge from the unconfined aquifer is by wells, evapotranspiration, and seepage to streams.

The principal source of recharge to the confined aquifer is seepage from mountain streams that flow across the alluvial fans flanking the valley floor. The blue clay series is less continuous along the edge of the valley, permitting recharge to beds that constitute the confined aquifer in the main part of the valley. The mountain streams show significant losses as they cross the porous surface of the fans, many reduced to no surface flow after a few miles.

The confined aquifer underlies almost the entire valley, and the water has sufficient head to flow at the land surface in an area of approximately 1430 square miles. The major discharge from the confined aquifer is by wells, springs, and upward leakage through the confining beds into the unconfined aquifer. A small amount may discharge as underflow into New Mexico.

Quality of water in the confined aquifer generally is better than that in the unconfined aquifer. The concentration of dissolved solids in 41 samples from the confined aquifer ranged from 70 to 437 mg/1 (milligrams per liter) and, in 271 samples from the unconfined aquifer, ranged from 52 to 13,800 mg/1. The least mineralized water in the unconfined aquifer occurs on the west side of the valley, where the Rio Grande provides irrigation water. The mineral concentration increases toward the sump area of the closed basin, probably because of solution of aquifer materials and by evaporative concentration in areas of a shallow water table.

PART I—STUDY OF GROUNDWATER RESOURCES OF EASTERN COLORADO

Refer to the *Ground Water Atlas of the United States, Segment 2—Arizona, Colorado, New Mexico, Utah—* Hydrologic Investigations Atlas 730-C. Go through pertinent sections of the atlas and summarize the characteristics of each of the three regional aquifer systems, using the outline provided.

Fill in the pertinent summary data for each aquifer system in Table 3.1.

Age, lithology, thickness, and properties of hydrogeologic units

Flow regime including nature of recharge and discharge, with water budget where available

Nature of potentiometric surface(s) and water-level conditions—confined and/or unconfined aquifers

Groundwater chemistry and quality

Well yields and groundwater withdrawals

Unique aspects and/or problems

Comparison of Aquifer Systems

Compare the sedimentary rock aquifers of the Denver Basin with the unconsolidated deposits of the High Plains. In each case, the aquifers are of approximately the same thickness. What differences can you ascribe to the effect of lithification?

Compare the aquifers of the High Plains with those of the Rio Grande (San Luis Valley). In each case, the aquifers are unconsolidated deposits. What differences can you ascribe to the effect of thickness?

PART II—HYDROGEOLOGY AND GROUNDWATER RESOURCES: PARKER, COLORADO

Douglas County is the fastest growing county in Colorado, and much of the growth is concentrated around Parker (Fig. 3.5; elevation 6000 ft). Future development will be limited by water supply, which, in this area, is mostly from groundwater.

You are asked to study this area, and, using the information contained in the *Ground Water Atlas of the United States*, prepare a report for the city of Parker that could be used to target an aquifer for a new municipal well. The RASA report for this area has not been published, but additional information on these aquifers is available in two

Table 3.1—REGIONAL AQUIFERS OF COLORADO SUMMARY CHARACTERISTICS			
Regional Aquifer System	Hydrogeologic Units Lithology, Characteristics, Depth, Thickness	Recharge, Discharge, and Groundwater Flow	Potentiometric Surface(s) Confined or Unconfined Water-Level Conditions
Denver Basin			
Colorado Portion of the High Plains Aquifer			
San Luis Valley Portion of the Rio Grande Aquifer			

Table 3.1—REGIONAL AQUIFERS OF COLORADO SUMMARY CHARACTERISTICS (continued)			
Unique Aspects and/or Problems	Groundwater Quality	Well Yields and Groundwater Withdrawals	Regional Aquifer System
			Denver Basin
			High Plains Aquifer
			Rio Grande Aquifer

water-supply papers: *Bedrock Aquifers of the Denver Basin, Colorado—A Qualitative Water-Resources Appraisal* (Robson, 1987) and *Alluvial and Bedrock Aquifers of the Denver Basin—Eastern Colorado's Dual Groundwater Resource* (Robson, 1989).

This report should address the total groundwater resources of Parker, describing potential aquifers in sufficient detail so that the city engineer can make an informed recommendation to the city council. Conclude the report with your recommendation for a particular aquifer, stating your reasons.

An outline for the report is given here.

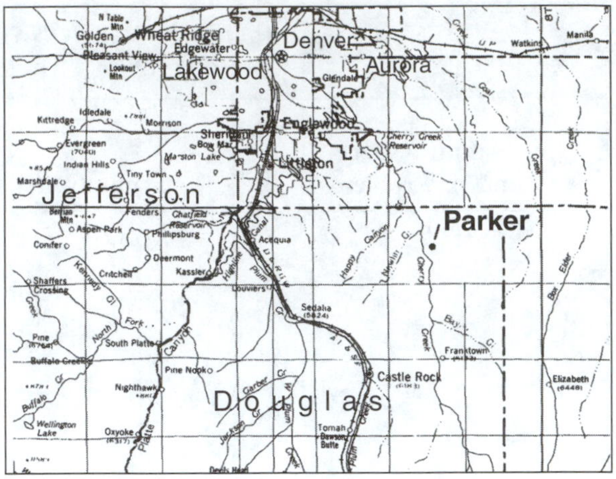

Figure 3.5—Location of Parker, CO.

GROUNDWATER RESOURCES AT PARKER, DOUGLAS COUNTY, COLORADO INCLUDING RECOMMENDATION FOR NEW OR ADDITIONAL AQUIFER DEVELOPMENT

- Potential groundwater sources
  - Surface alluvial aquifer
  - Bedrock aquifers
    - Lithology, thickness, depth, and properties
    - Recharge and discharge
    - Water chemistry and quality
    - Well yields
    - Current use and groundwater withdrawals
    - Water levels
    - Applicable water law/ownership of groundwater
    - Other considerations
- Recommendation for new or additional aquifer development
  - Recommended aquifer
  - Rationale for selecting aquifer
  - Predicted effects of pumping

LAB 4

REGIONAL AQUIFER STUDY: GULF COASTAL LOWLANDS AQUIFER

PURPOSE: Introduce you to the scope and magnitude of regional aquifers and their systems. Introduce you to groundwater occurrences in diverse aquifer materials under varying conditions. Introduce you to the data resources of *Ground Water Atlas of the United States* and to the RASA program.

OBJECTIVES: Determine the regional aquifers in a given study area, and learn about their basic properties. Investigate one aquifer in substantial detail, sufficient to site a municipal well at, for example, Houston or Baton Rouge.

EXERCISE OVERVIEW: You will read a brief summary of the coastal lowlands regional aquifer system provided below and outline the characteristics of the aquifer system using the *Ground Water Atlas*. You will then study one area in detail to provide the background information used in selecting an aquifer for a municipal well, using the atlas and one of the RASA studies published by the U.S. Geological Survey. Major cities in this area include Houston, TX, Baton Rouge, LA, New Orleans, LA, and Mobile, AL.

Lab 3 deals with a study area in Colorado. Your instructor may ask you to study this area also, or, in fact, any other area in the world where sufficient information on regional aquifers has been published. The procedures and questions used in this exercise can be used as a template.

MATERIALS REQUIRED: *Ground Water Atlas of the United States, Segment 5* (*Arkansas, Louisiana, Mississippi*) (Renken, 1998)
Summary of Hydrology of the Regional Aquifer Systems, Gulf Coastal Plain, South-central United States, USGSPP 1416-A (Grubb, 1998)
Geohydrologic Units of the Coastal Lowlands Aquifer System, USGSPP 1416-C (Weiss, 1992)

STUDY AREA: The Gulf Coastal Lowlands Aquifer (GCLA) extends from Mexico to Florida in a coastal belt about 80 miles wide, with the aquifer system extending offshore about 100 miles to the edge of the continental shelf (Fig. 4.1).The Vicksburg–Jackson regional confining unit separates the GCLA from the two adjacent regional aquifers, the Texas Coastal Uplands Aquifer System and the Mississippi Embayment Aquifer System (Fig. 4.1).

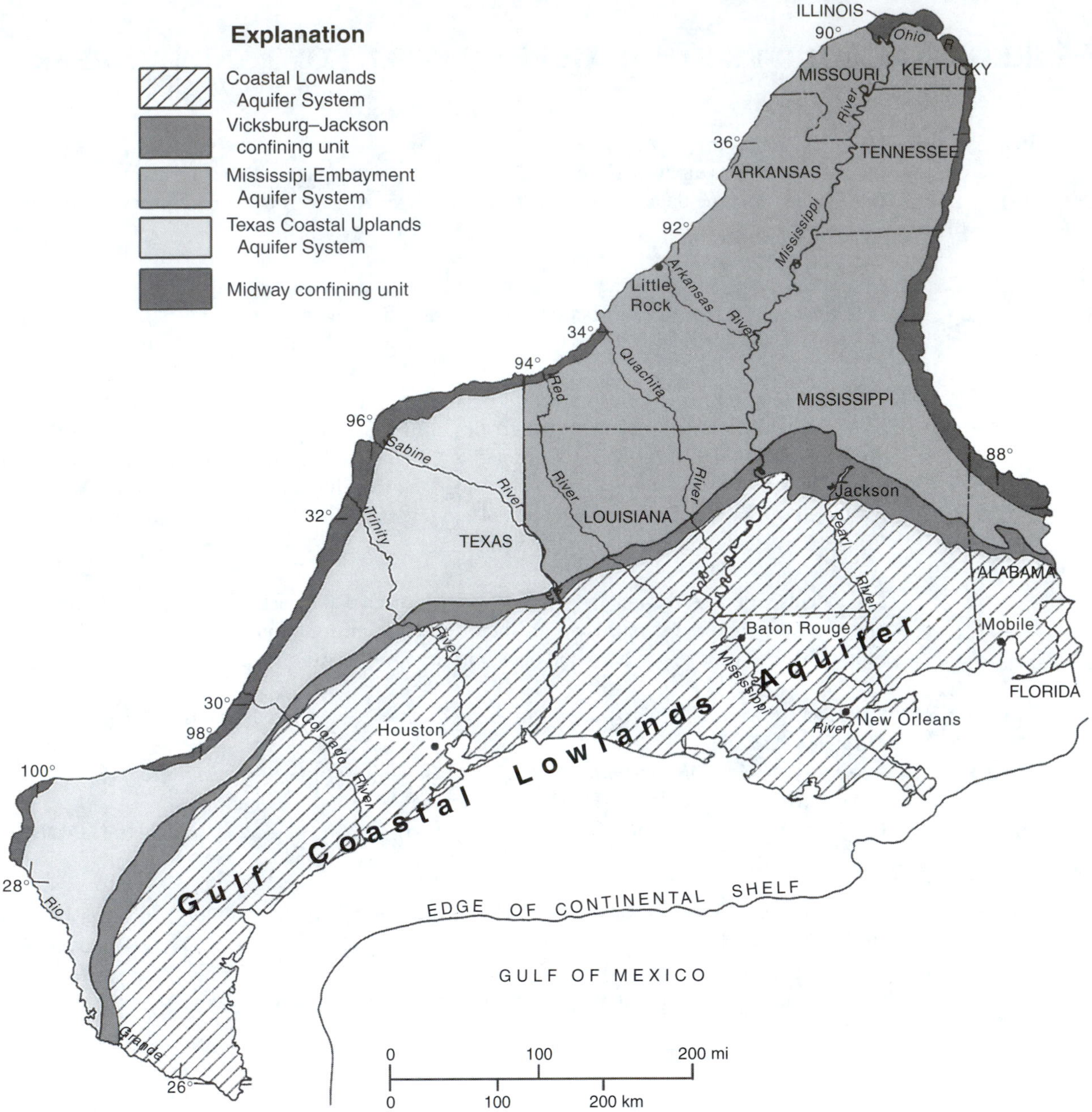

Figure 4.1—Regional aquifer systems and confining units in the Gulf Coast area (from Weiss, 1992, Figure 2).

Characteristics of the Gulf Coast Lowland Aquifer System[1]

The coastal lowlands aquifer system consists of sediments of Oligocene age and younger. The sediments are predominantly interbedded sands, silts, and clays, with minor amounts of lignite and limestone. Average thickness of the sediments is about 6000 feet, with a maximum thickness of more than 18,000 feet occurring offshore from southern Louisiana.

The base of the coastal lowlands aquifer system is the top of the Vicksburg–Jackson confining unit, which is a massive clay that represents the last major transgression of the sea. A zone of abnormally high fluid pressure (geopressured zone) is present above the top of the Vicksburg–Jackson confining unit onshore in a narrow band along the coast of Texas and Louisiana and on the continental shelf. Where the geopressured zone is present, it is considered to be the base of the coastal lowlands aquifer system. This relationship is shown in Figure 4.2

The sediments in the coastal lowlands aquifer system are divided into five permeable zones (7–11 in Fig. 4.2) and two confining units (16 and 17 in Fig. 4.2). The permeable zones are not separated by intervening, regionally mappable confining units in most of the study area.

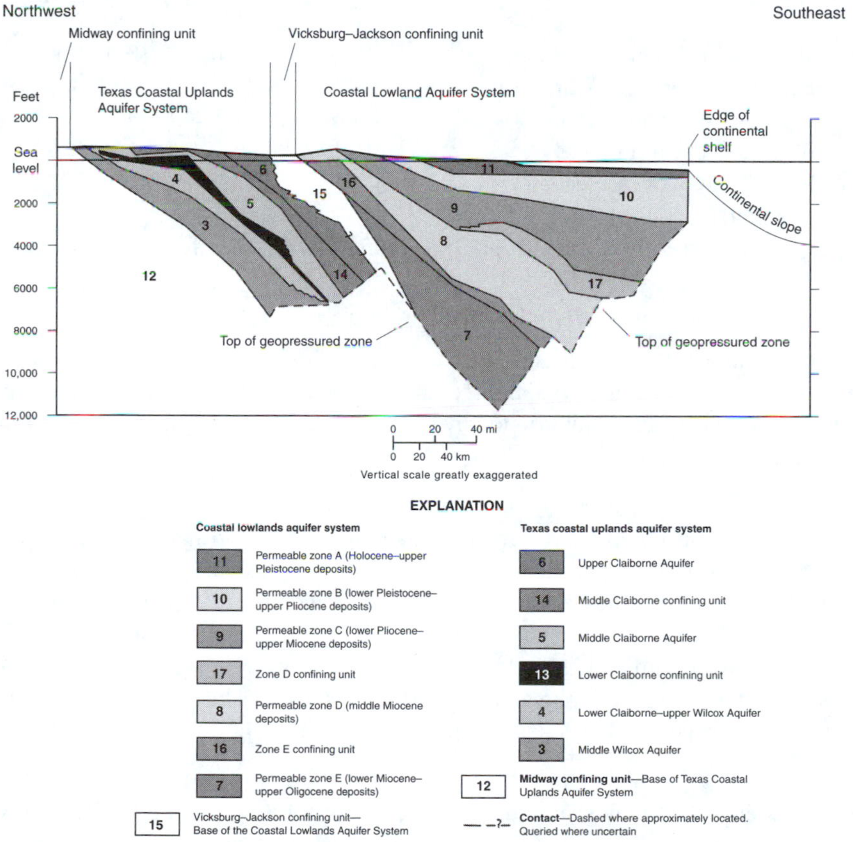

Figure 4.2—Schematic cross section across the GCLA (from Grubb, 1998, Figure 9).

[1]Abstracted from Weiss, 1992, and Renken, 1998.

Average sand percentage of the permeable zones ranges from about 40 percent to more than 60 percent. However, the areal distribution of sand is variable within and among permeable zones. A lobate pattern of greater sand percentages is typical of the permeable zones, and all zones except one have at least one area with sand percentage greater than 80 percent.

Water chemistry in these aquifers follows the Chebotarev sequence, being primarily calcium bicarbonate waters near the outcrop, progressing down-dip to sodium bicarbonate waters, and eventually becoming sodium chloride at depth. Total dissolved solids (TDS) increase with depth, as well.

Several major rivers cross the GCLA, such as the Rio Grande, Trinity, Sabine, Mississippi, and Pearl Rivers (Fig. 4.1). To varying degrees, these rivers have entrenched their courses into the gently dipping aquifers of the GCLA, and the rivers' alluvial aquifers hydraulically connect with the regional aquifers, as shown schematically in Figure 4.3.

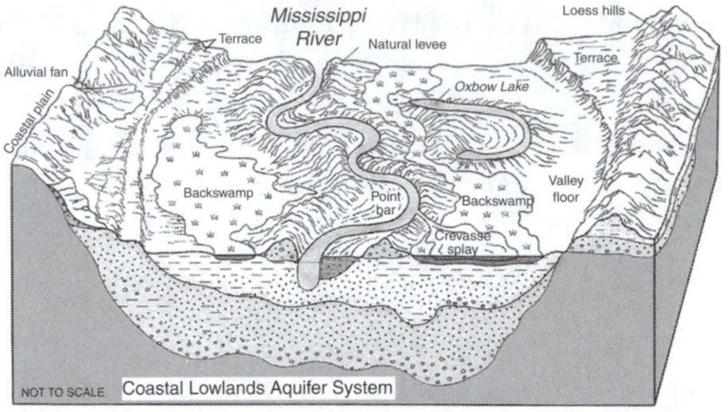

Figure 4.3—The Mississippi River cut into the coastal plain sediments during pluvial episodes of the Pleistocene and deposited varied alluvial materials (from Renken, 1998, Figure 24).

The GCLA system consists largely of sediments deposited in a deltaic to marginal marine environment, in which changes in lithologic facies are rapid, numerous, and complex. An aquifer's gulfward extent is determined, in part, by progressive facies changes as permeable deltaic sands grade seaward to less permeable prodelta silts and clays. These prodelta sediments are fine-grained terrigenous clastics deposited from suspension seaward of the delta front. The aquifer system, therefore, contains a highly interbedded mix of sand and clay.

Thick sand beds of wide areal extent are uncommon, and the lack of widespread clay beds means that few continuous confining units can be used to divide the section into the customary sequence of alternating aquifers and confining units. Aquifer continuity is also interrupted by extensive shoreline-striking, gulfward-dipping growth faulting that has produced numerous listric faults.

Many local aquifer names have been applied to parts of the aquifer system, primarily in Louisiana. The *Chicot aquifer* and the *Evangeline aquifer* are names used in southwestern Louisiana, whereas in the industrial districts of Baton Rouge and New Orleans, simple descriptive names are used, like the *400-foot sand* and the *1200-foot sand*, which refer to their depth. At Baton Rouge, useful sands occur every few hundred feet from 400 ft to 2800 ft. Because of the regional southward dip and listric faults, the *1200-foot sand* at New Orleans is not the same aquifer as the *1200-foot sand* at Baton Rouge. Despite all these difficulties, studies have indicated that the aquifer system can be divided usefully into five permeable zones of regional extent (Fig. 4.2).

The coastal lowlands aquifer system yields large quantities of water for agricultural, public supply, domestic, and industrial use. Large-capacity industrial and municipal wells flow as much as 4000 gallons per day when they are first drilled. Many large-capacity wells screened opposite water-bearing sands of Permeable Zones A–D typically yield more than 1000 gallons per minute.

Because the aquifers are unconsolidated sediments, groundwater withdrawals can and do induce land surface subsidence. In inland areas, this poses little problem, but in Houston and New Orleans the subsidence actually lowers some areas below sea level. The shallower aquifers in this system are also susceptible to saltwater encroachment where heavily pumped.

PART I—STUDY OF GROUNDWATER RESOURCES OF THE GULF COAST

Refer to the *Ground Water Atlas of the United States, Segment 5—Arkansas, Louisiana, and Mississippi—* Hydrologic Investigations Atlas 730-F. Go through pertinent sections of the atlas, and summarize the characteristics of each of the regional aquifers, using the outline provided.

Fill in the pertinent summary data for each of the regional aquifers in Table 4.1.

Age, lithology, thickness, and properties of hydrogeologic units

Flow regime including nature of recharge and discharge, with water budget where available

Nature of potentiometric surfaces and water-level conditions—confined and/or unconfined aquifers

Groundwater chemistry and quality

Well yields and groundwater withdrawals

Unique aspects and/or problems

Comparison of Aquifer Systems

Compare the yields of confined, sedimentary rock aquifers of the Denver Basin, Lab 3, with the confined, unconsolidated sediment aquifers of the Gulf Coast. In each case, the aquifers are of comparable depth and thickness. What differences can you ascribe to the effect of lithification? (If you haven't worked the Colorado exercise, Lab 3, try to obtain data from some local area where the aquifers are primarily sandstones in an artesian basin. If no data are available, give the question a try simply using your knowledge and logic.)

PART II—HYDROGEOLOGY AND GROUNDWATER RESOURCES: BATON ROUGE, LOUISIANA

Using the information contained in the *Ground Water Atlas of the United States*, prepare a report for the city of Baton Rouge (or some other area designated by your instructor) that could be used to target an aquifer for a new municipal well. The RASA report for this area is published as U.S. Geological Survey Professional Paper 1416 (in several parts—1416-A, 1416-B, etc.).

This report should address the total groundwater resources of Baton Rouge, describing potential aquifers in sufficient detail so that the city engineer can make an informed recommendation to the city council. Conclude the report with your recommendation for a particular aquifer, stating your reasons.

A useful supplement to the RASA report that provides more detailed information for the Baton Rouge area is the water-supply paper by Meyer and Turcan, 1955.

An outline for the report is given here.

Table 4.1—REGIONAL AQUIFERS OF THE GULF COAST LOWLAND AQUIFER SYSTEM SUMMARY CHARACTERISTICS			
Regional Aquifer	**Hydrogeologic Units Lithology, Characteristics, Depth, Thickness**	**Recharge, Discharge, and Groundwater Flow**	**Potentiometric Surface(s) Confined or Unconfined Water-Level Conditions**
Permeable Zone A			
Permeable Zone B			
Permeable Zone C			

Table 4.1—REGIONAL AQUIFERS OF THE GULF COAST LOWLAND AQUIFER SYSTEM SUMMARY CHARACTERISTICS (continued)

Unique Aspects and/or Problems	Groundwater Quality	Well Yields and Groundwater Withdrawals	Regional Aquifer
			Permeable Zone A
			Permeable Zone B
			Permeable Zone C

GROUNDWATER RESOURCES AT BATON ROUGE
EAST BATON ROUGE PARISH, LOUISIANA
INCLUDING RECOMMENDATION FOR NEW OR ADDITIONAL AQUIFER DEVELOPMENT

- Potential groundwater sources
    - Surface alluvial aquifer
    - Deeper aquifers
        - Lithology, thickness, depth, and properties
        - Recharge and discharge
        - Water chemistry and quality
        - Well yields
        - Current use and groundwater withdrawals
        - Water levels
        - Applicable water law/ownership of groundwater
        - Other considerations
- Recommendation for new or additional aquifer development
    - Recommended aquifer
    - Rationale for selecting aquifer
    - Predicted effects of pumping

LAB 5

WATER CHEMISTRY AND WATER QUALITY

PURPOSE: Interpret water analysis data for determining groundwater flow and for water quality. Relate water chemistry to geology.

OBJECTIVES: Relate water chemistry and water quality to geology using groundwater maps of the Colorado Front Range Urban Corridor.

Interpret analytical data from well samples in the Amargosa Desert of Nevada–California to estimate regional groundwater flow.

Evaluate suitability of groundwater for domestic and agricultural use.

INTRODUCTION

More than 99 percent of the dissolved solids in natural groundwater is composed of fewer than a dozen constituents. The bulk of these consist of only seven ions (Ca^{2+}, Na^{+}, Mg^{2+}, K^{+}, HCO_3^{-}, SO_4^{2-}, and Cl^{-}) and some dissolved SiO_2 (Table 5.1). Routine chemical analyses of water samples report these constituents, so they can be used to provide information both on the aquifer materials and the suitability of a water for human use, often referred to as *water quality*.

Table 5.1—CHEMISTRY OF NATURAL WATERS

Constituent (mg/l)	Precipitation[1]	Groundwater in Unconsolidated Deposits[2]	Groundwater in Igneous Rocks[3]	Groundwater in Sedimentary Rocks[4]	Groundwater in Carbonates[5]	Seawater[6]
Na	0.6	47	4	20	13	10560
K	0.4	3	1	2	3	380
Ca	0.9	54	8	53	55	400
Mg	0.2	15	2	19	28	1272
HCO_3	2.0	157	40	263	255	142
SO_4	3.0	64	1	47	48	2560
Cl	0.4	21	1	12	14	18980
NO_3	0.3	0.6	n/a	2.7	n/a	<1
SiO_2	0.1	22	19	15	n/a	<1–4[7]
TDS	5.1	230	76	380	416	34378
pH	5.5	7.5	6.8	7.5	7.5	8.1–8.4[8]

[1] Freeze and Cherry, 1979: Median values of seven sites worldwide; hundreds of analyses.
[2] Hem, J.D., 1970
[3] Freeze and Cherry, 1979: 191 samples at 90 localities; median values of 9 means.
[4] Hem, 1970: Average of six samples.
[5] Freeze and Cherry, 1979: Mean of four means.
[6] Hem, 1959.
[7] Varies biochemically; average about 4 ppm.
[8] Varies; surface—8.1–8.4; seawater pond with active photosynthesis—8.6; Black Sea at 1000 m with H_2S—7.26.

The general public often assumes that all pristine groundwaters are of high quality, suitable for all intended uses. Based on major ion chemistry alone, you may see that this is not always true.

Many problems of groundwater contamination are caused by constituents in trace amounts. In this lab, you will work only with major ions and a few minor ions (Table 5.2). One of these, however, NO_3^-, is useful as an indicator of contamination because it normally doesn't occur naturally at concentrations above a few ppm[1]. Thus a reported concentration of, for example, 10 ppm NO_3^-, while below drinking-water standards, would flag the water as probably contaminated, which should lead to other tests, such as a test for fecal coliform bacteria.

Table 5.2—DRINKING-WATER STANDARDS

Major Ions[1]		
Constituent	**Concentration (mg/l)**	**Comment**
SiO_2	—	SiO_2 not ionic—probably H_4SiO_4 Boiler, heat exchanger problems
Na	20 (WHO-200)	Irrigation problems common
K	—	
Ca	200	Main constituent of hardness
Mg	125	A constituent of hardness
HCO_3, CO_3	500	
SO_4	250	
Cl	250	
NO_3	44 (10 as NO_3–N]	Indicator of organics and/or fertilizers
TDS	500	
pH	6.5–8.5	
Minor Constituents—Maximum Contaminant Levels[2]		
Zn (5), F (4), Ba (2), Cu (1.3), H_2S (1)		H_2S—foul odor; F–teeth mottling
Fe (0.3), CN (0.2), Ag, Cr, Ni (0.1)		Fe—stain
Al, As, Mn, Se (.05) (50 μg/l)		Se in Cretaceous shales, locoweed
Pb (15 μg/l), Cd (5 μg/l), Hg (2 μg/l)		
U (20 μg/l; 30 pCi/1) *Escherichia coli* (0), *Giardia lamblia* (0)		Any *E. coli* or *Giardia lamblia* renders water unfit

[1]E.P.A. National Secondary Drinking Water standards; these are non-enforceable guidelines (July 2001).
[2]Maximum contaminant levels are enforceable standards for public water supplies. See EPA for current standards of trace elements, organic compounds, and radionuclides at http://www.epa.gov/safewater/mcl.html (July 2001).

---

[1]ppm, parts per million; water analyses are expressed in units of both ppm (weight/weight basis) and mg/l (mass/volume basis); but at normal groundwater salinities, the differences are negligible (up to several thousand ppm/mg/l).

GROUNDWATER MAPS—FRONT RANGE URBAN CORRIDOR

Introduction[2]

Knowledge of the well yields and chemical quality of water from water-table aquifers may assist state and local officials in making decisions regarding land-use conversions and in locating water supplies in the rapidly urbanizing Front Range Urban Corridor (Fig. 5.1). In this corridor, the principal water-table aquifers consist of thick, unconsolidated alluvial deposits that are perennially saturated. These deposits occur in stream valleys and in terraces, both along present stream valleys and on slopes of the foothills east of the Front Range. Locally, saturated permeable materials also occur in the upper weathered and fractured zone of consolidated sedimentary rocks; and in the crystalline rocks that form the Front Range in the western part of the area, water occurs locally in fractures.

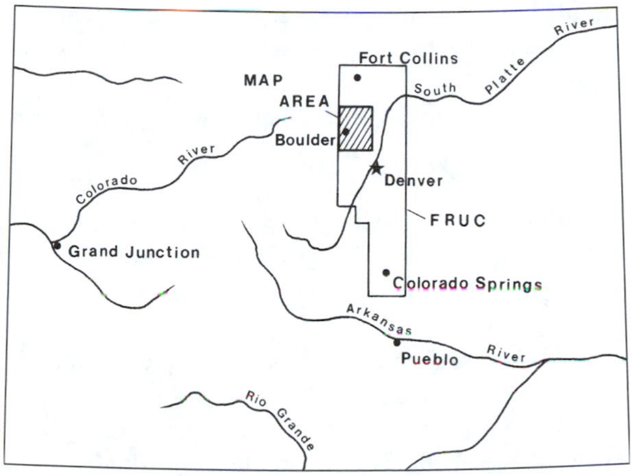

Figure 5.1—Index map of Colorado showing the Front Range Urban Corridor and area of map shown in Figure 5.2 (from Hillier and Schneider, 1979).

The concentration of dissolved solids was the principal criterion used in this investigation to determine the suitability of groundwater for urban development. Water containing 500 mg/l or less of dissolved solids generally is suitable for uses associated with urban development. However, excessive concentrations of individual dissolved constituents, values of selected physical properties, amounts of radioactivity, and numbers of fecal coliform bacteria in the water may cause the water to be unsuitable for a particular use.

The dissolved-solids concentrations of water from water-table aquifers are shown on the map in Figure 5.2. Also shown are individual dissolved constituents that exceeded state standards for public-water supplies.

The taste of water is affected by dissolved solids, chloride, iron, and manganese. Dissolved solids and chloride impart a salty taste to the water; iron and manganese impart a bitter metallic taste to the water and to beverages made using the water. Discolored water, stained laundry and porcelain fixtures, and encrusted plumbing are caused by iron and manganese. Encrusted plumbing also is caused by dissolved solids and excessive hardness. In addition, excessive hardness may reduce the useful life of hot-water heaters and impairs the quality of frozen or canned fruits and vegetables. Drinking water containing excessive magnesium and sulfate may have a laxative effect on people who are unaccustomed to drinking the water. However, this condition abates with continued consumption of the water.

Fluoride, nitrite plus nitrate, and selenium in water may be health hazards. While fluoride is known to reduce dental cavities, concentrations greater than 1.8 mg/l may cause mottling of teeth, especially in children. Concentrations of nitrite plus nitrate as nitrogen greater than 10 mg/l may cause methemoglobinemia (blue-baby disease) in infants less than 9 months old who drink the water or who are breast-fed by mothers who drink the water. Concentrations of nitrite plus nitrate as nitrogen in the study area greater than 10 mg/l usually indicate contamination from septic-tank systems, barnyards, corrals, or commercial fertilizers. Concentrations of selenium greater than 10 μg/l (0.010 mg/l) may cause selenium poisoning in people and livestock. Alkali disease that afflicts livestock is caused by selenium.

---

[2] This introductory text on the Front Range Urban Corridor is from Hillier and Schneider, 1979.

All parameters that govern the suitability of water for a public supply were not determined during this investigation. In addition to the chemical constituents included in this report, concentrations of trace elements, as well as levels of certain pesticides, radioactivity, fecal coliform bacteria, ammonia, color, foaming agents, odor, hydrogen sulfide, and turbidity could affect the suitability of the water for various uses.

Procedure

Examine the map in Figure 5.2, which shows generalized hydrogeologic units, wells, and total dissolved solids (TDS) content of well samples.

1. Characterize the general quality of groundwater in the Precambrian crystalline rocks of the Front Range.

2. What are the individual problem constituents, if any, in the waters in the Precambrian crystalline rocks—that is, what would cause them to be rejected for a public water supply?

3. Characterize the general quality of groundwater in the unconfined sedimentary rock aquifers.

4. What are the individual problem constituents in the unconfined sedimentary rock aquifers, if any?

5. The alluvial deposits have been subdivided into two map units—those alluvial aquifers that have TDS of less than 500 mg/l and those with greater than 500 mg/l. The map shows that alluvial wells have TDS values that actually range from 53 mg/l to more than 3000 mg/l. Why do you think the division was made at 500 mg/l?

6. Characterize the general quality of groundwater in the alluvium that has less than 500 mg/l TDS. What are the individual problem constituents, if any?

7. Characterize the general quality of groundwater in the alluvium that has greater than 500 mg/l TDS. What are the individual problem constituents, if any?

8. Summarize the changes in groundwater that occur in the alluvium downstream from the mountains.

9. Note that, although most of the alluvial deposits are restricted to fluvial valleys, some, like those at Rocky Flats in the southwestern part of the map, are more widespread and are not associated with modern drainages. These are Pleistocene alluviums that veneer pediments. Do these pediment alluviums have substantial precipitation recharge?

10. Summarize the problem constituents in all of the unconfined aquifers.

11. Are the problem constituents natural or anthropogenic—that is, are the groundwaters unsuitable for public supply natural or contaminated by human activity?

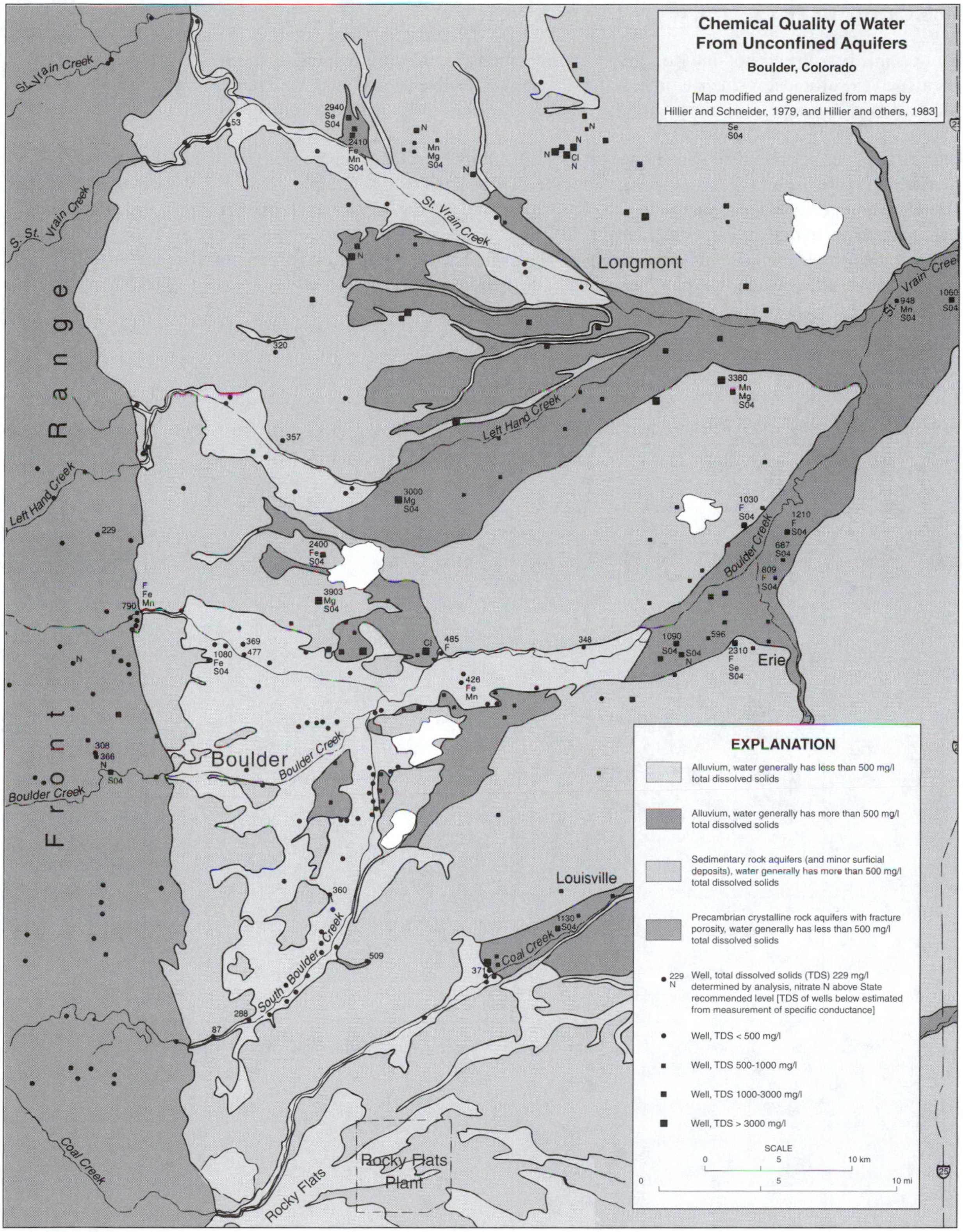
Chemical Quality of Water
From Unconfined Aquifers
Boulder, Colorado
[Map modified and generalized from maps by Hillier and Schneider, 1979, and Hillier and others, 1983]
EXPLANATION
Alluvium, water generally has less than 500 mg/l total dissolved solids
Alluvium, water generally has more than 500 mg/l total dissolved solids
Sedimentary rock aquifers (and minor surficial deposits), water generally has more than 500 mg/l total dissolved solids
Precambrian crystalline rock aquifers with fracture porosity, water generally has less than 500 mg/l total dissolved solids
229 N Well, total dissolved solids (TDS) 229 mg/l determined by analysis, nitrate N above State recommended level [TDS of wells below estimated from measurement of specific conductance]
Well, TDS < 500 mg/l
Well, TDS 500-1000 mg/l
Well, TDS 1000-3000 mg/l
Well, TDS > 3000 mg/l
SCALE
0 5 10 km
0 5 10 mi
Front Range
Longmont
Boulder
Louisville
Erie
Rocky Flats Plant
Rocky Flats
St. Vrain Creek
S. St. Vrain Creek
Left Hand Creek
Boulder Creek
South Boulder Creek
Coal Creek

Figure 5.2

WATER CHEMISTRY

Water chemistry is useful not only for quality considerations, but also for helping to understand groundwater flow systems; given sufficient residence time, water (the universal solvent) will react with porous media and its composition will change. In some areas, distinct chemical facies of groundwater are recognized.

Given large tables of analytical data, however, it is difficult to interpret differences in the chemical nature of the waters. Graphs are useful for this purpose, and several specialized types of graphs are in use. We will work with one of these diagrams, the Piper diagram, which plots cations and anions on separate ternary (trilinear) diagrams and then combines them onto a diamond-shaped field. These diagrams allow us not only to see graphically the nature of a given water sample, but also the relationship to other samples (Fig. 5.3). If two different waters are mixed, for example, the resulting water will plot on a straight line between the parent waters.

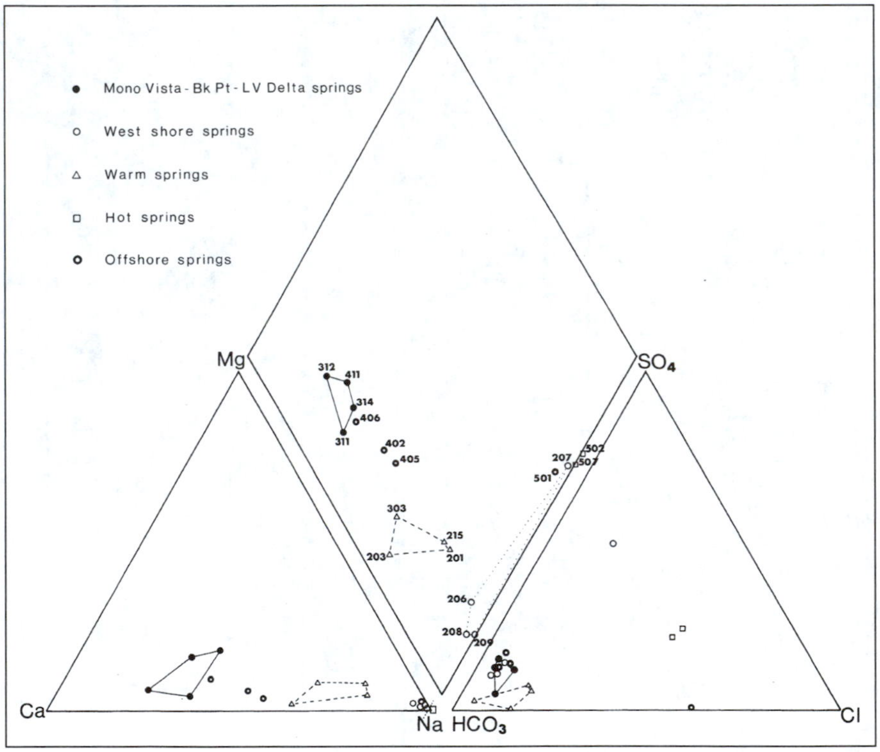

Figure 5.3—Composition of shoreline and offshore springs, Mono Lake, California (Lee, 1969).

To show the chemical nature of waters on a Piper diagram, constituents are expressed in terms of percentage of reacting values, where reacting values are, in turn, expressed in equivalents per million (epm) or milligram equivalents per liter. The relationship is

Concentration (epm) = concentration (ppm) × valence/molecular weight

For example, look at the analysis from the first well in Table 5.3 on page 43. Mg^{2+} is present at 21 ppm:

$$Mg^{2+} \text{ concentration (epm)} = 21 \times 2/24.31 = 1.7 \text{ epm}$$

[*Note: The box on the right contains an abbreviated table of weights and valences; you will need access to a periodic table of the elements to determine molecular weights of any other constituents.*]

PARTIAL PERIODIC TABLE QUICK REFERENCE/GROUNDWATER		
Element	**Weight**	**Valence**
H	1.01	1+
C	12.01	4+
N	14.01	5+
O	16.00	2−
F	19.00	1−
Na	22.99	1+
Mg	24.31	2+
Si	28.09	4+
S	32.06	6+
Cl	35.45	1−
K	39.10	1+
Ca	40.08	2+
Fe	55.85	2+, 3+

When all the cations have been expressed as equivalents per million, the percentage of each is determined. Again, from the first well analysis, the sum of the cations is 8.1 epm, so Mg^{2+} is 1.7/8.1, or 21 percent, of the cations.

To plot the analytical results, start with the cation ternary diagram. As shown in Figure 5.4, a ternary diagram has three axes, along which the values of each cation are plotted. (Because there are only three variables, Na^+ and K^+ are combined at one apex.) Using the data from the first well, Mg^{2+} constitutes 21 percent of the cations, so the data point will fall somewhere along the horizontal line indicating 21 percent Mg^{2+} (upper left diagram).

Similarly, Ca^{2+} is 32 percent and will plot along the 32 percent Ca^{2+} line, as shown in the upper right diagram. When combined, the data point will fall where the two lines intersect (lower diagram). To check the accuracy of your plotting, you can read the percentage of the third variable, $Na^+ + K^+$, and it should agree with the tabulated value, 47 percent.

If a cation other than these is present in significant amounts, it must be combined with one of the apical constituents, or the percentage reacting values must be calculated on a basis excluding the additional cation.

Anions are plotted in the same way. The two data points are then combined into a diamond-shaped field that shows the total chemical nature of the water (Fig. 5.5).

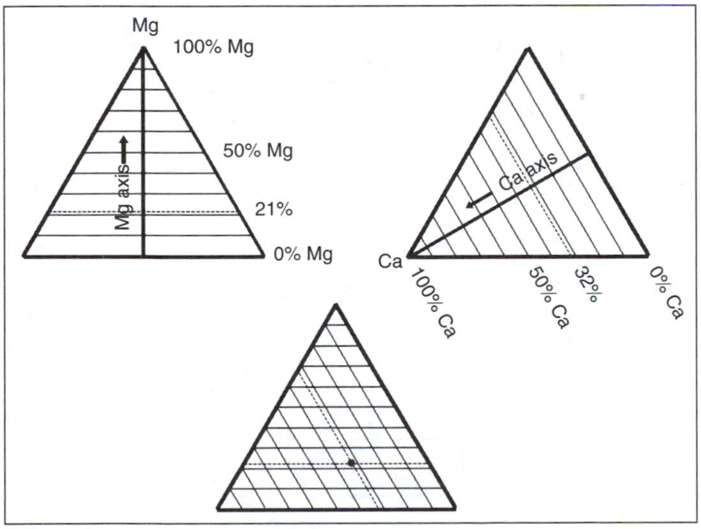

Figure 5.4—Using ternary diagrams to plot cations.

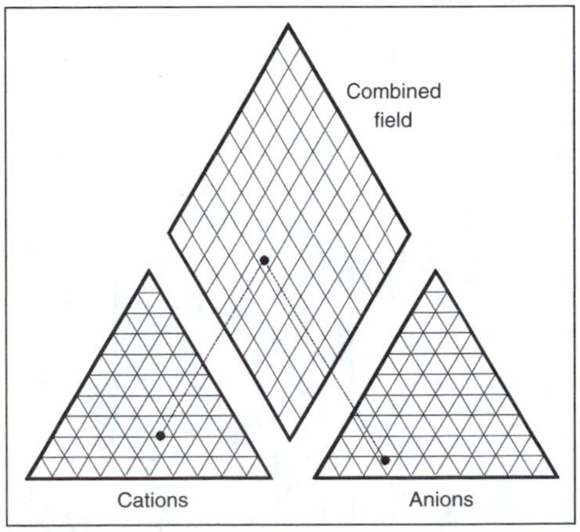

Figure 5.5—Plotting analyses on a Piper diagram.

Because the total equivalent weight of cations must equal the total equivalent weight of anions in a water, the sum of the cations (in epm) should equal the sum of the anions (in epm). Any difference is a measure of the error in the analysis, and can be expressed by

$$\text{Error} = \frac{\text{cations (epm)} - \text{anions (epm)}}{\text{cations (epm)} + \text{anions (epm)}} \times 100\%$$

WATER CHEMISTRY—AMARGOSA DESERT

The Nevada Test Site is in southern Nevada, about 50 miles northwest of Las Vegas (Fig. 5.6). The Amargosa Desert is a large valley just to the south, and regional groundwater flow from the test site may pass under the Amargosa. The area is in the Basin and Range physiographic province, where groundwater flow systems are sometimes complex.

Climate in the Amargosa is arid, typical of the Sonora Desert. Precipitation varies from more than ten inches in the mountain ranges to less than four inches in the valley.

Geology is typical of the Basin and Range; block-faulted mountain ranges shed detritus into elongate basins, often accumulating thousands of feet of alluvium. Although the hydrogeologic framework is complex, it can be generalized as follows: unconsolidated basin-fill sediments at the surface hold unconfined water in closed topographic basins; a deep, Paleozoic carbonate aquifer carries regional flow for long distances under the topo basins; and semi-confining ash-flow tuffs separate the aquifers.

Water analyses are provided from 15 samples in the Amargosa Desert (Table 5.3). Interpret these data to determine if there is more than one groundwater flow system.

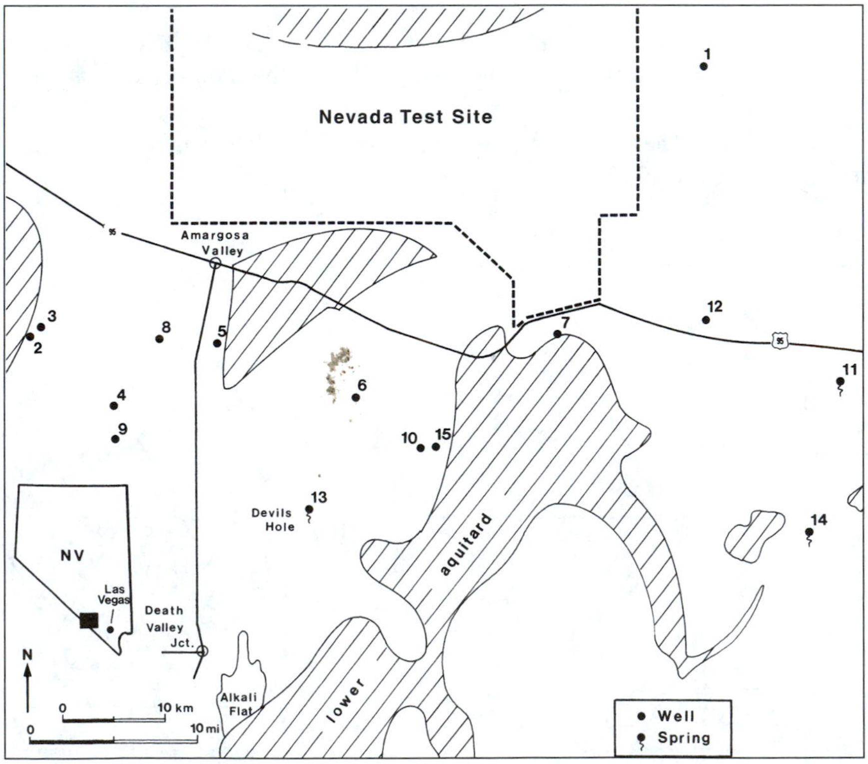

Figure 5.6—Sample locations, Amargosa Desert (modified from Winograd and Thordarson, 1975).

1. Most of the analytical data have been reduced for you, but you will have to reduce data from the last two wells, numbers 14 and 15. These data are provided for you in LAB5DATA.xls on the CD. Reduce one of the two samples manually (with your calculator) in order to reinforce your understanding of the methods. Once you have completed the data reduction and checked for accuracy by calculating error, then run the data for the two samples with the spreadsheet to check your calculation and to reduce the data from the last sample.

2. Plot these analyses on the Piper diagram of Figure 5.7 for subsequent interpretation. Ten of the analyses have been plotted already; the last five have been left for you to plot.

3. Based on your Piper diagram, does there appear to be one flow system or more? If more than one type appears in the Piper plot, characterize each water type.

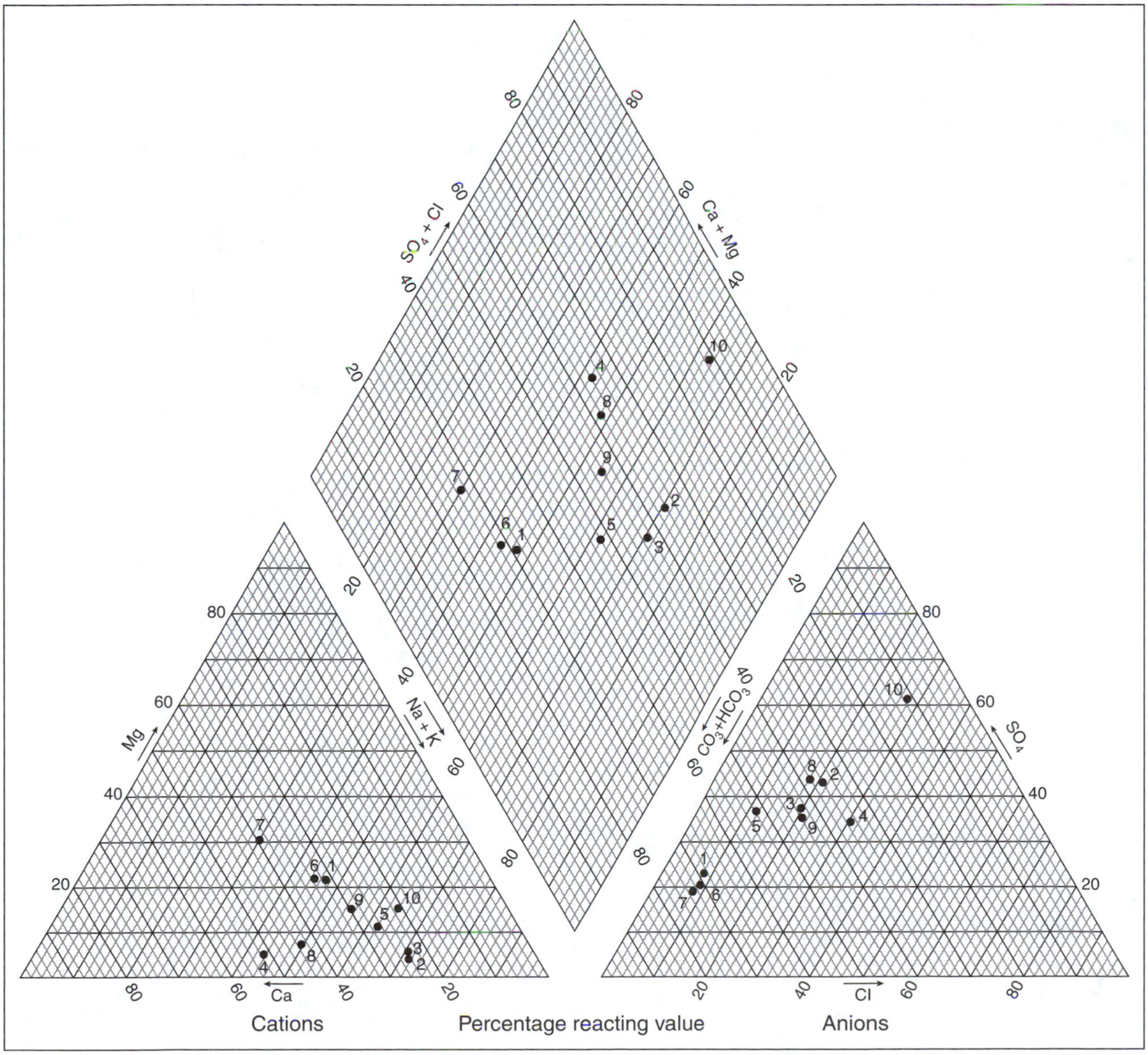

Figure 5.7—Piper diagram for plotting major-ion water chemistry of the Amargosa Desert.

4. Figure 5.6 shows the location of each of the samples. Using this map, along with the above information, describe, in general terms, the groundwater system(s) in the Amargosa.

5. You may refer to the case study on regional flow in the Great Basin in Fetter, 2001, pp. 250–254. The southern end of the cross section in Figure 7.15 in that textbook is the Amargosa Desert.

WATER QUALITY—AMARGOSA DESERT

Comment on the suitability of each of the wells in the Amargosa Desert for both domestic use and irrigation.

If any of the wells would be unsuitable for either purpose, indicate the reason(s) why. Domestic use should take into account drinking-water standards (Table 5.2), hardness, and any indication of possible contamination. Agricultural use should consider total dissolved solids and possible sodium problems (you might want to review these in your textbook, especially the sodium adsorption ratio [e.g., Fetter, 2001, p. 367]).

Table 5.3—WATER SAMPLE ANALYSES, AMARGOSA DESERT

WELL		Na			K			Ca			Mg			Cations		HCO_3		
No.	ID	ppm	epm	%	ppm	epm	%	ppm	epm	%	ppm	epm	%	ppm	epm	ppm	epm	%
1	TW-3	83	3.6	44.7	7.6	0.2	2.4	51	2.5	31.5	21	1.7	21.4	163	8.1	328	5.4	69.2
2	Finley	151	6.6	68.3	8.2	0.2	2.2	49	2.4	25.4	4.8	0.4	4.1	213	9.6	207	3.4	35.8
3	Defir	181	7.9	67.8	13	0.3	2.9	58	2.9	24.9	6.3	0.5	4.5	258	12	296	4.9	42.8
4	Smith	62	2.7	40.0	9.0	0.2	3.4	70	3.5	51.8	3.9	0.3	4.8	145	6.7	142	2.3	35.5
5	Cook	132	5.7	59.4	10	0.3	2.6	52	2.6	26.9	13	1.1	11.1	207	9.7	293	4.8	52.3
6	USGS	62	2.7	40.7	7.8	0.2	3.0	45	2.2	33.9	18	1.5	22.4	133	6.6	284	4.7	70.7
7	Army	37	1.6	27.7	5.2	0.1	2.3	47	2.3	40.3	21	1.7	29.7	110	5.8	256	4.2	72.7
8	Selbach	61	2.7	45.7	7.8	0.2	3.4	51	2.5	43.9	4.9	0.4	7.0	125	5.8	134	2.2	38.3
9	17/49	160	7.0	52.1	15	0.4	2.9	81	4.0	30.3	24	2.0	14.8	280	13	370	6.1	44.1
10	Rasmus	667	29	63.5	14	0.4	0.8	188	9.4	20.5	84	6.9	15.1	953	46	302	4.9	11.1
11	IndSpg	4.5	0.2	4.5	1.1	0.0	0.6	50	2.5	57.2	20	1.6	37.7	76	4.4	238	3.9	89.4
12	TW-10	7.6	0.3	8.7	1.0	0.0	0.7	41	2.0	53.8	17	1.4	36.8	67	3.8	200	3.3	87.7
13	D. Hole	67	2.9	40.0	8.1	0.2	2.8	52	2.6	35.6	19	1.6	21.5	146	7.3	306	5.0	69.0
14	CC Spg	1.8			0.4			73			12					283		
15	17/52	61			7.2			34			22					274		

WELL		Cl			SO_4			NO_3			Anions		TDS	Error	S.C.	pH
No.	ID	ppm	epm	%	ppm	epm	%	ppm	epm	%	ppm	epm	ppm	%		
1	TW-3	23	0.6	8.3	84	1.7	22.5				435	7.8	431	1.9	520	8.7
2	Finlay	71	2.0	21.1	196	4.1	43.1				474	9.5	582	0.7	930	7.9
3	Defir	80	2.3	19.9	203	4.2	37.3				579	11.3	687	1.2	1160	7.9
4	Smith	61	1.7	26.3	107	2.2	34.0	17	0.3	4.2	327	6.5	400	1.4	700	7.9
5	Cook	36	1.0	11.0	162	3.4	36.7				491	9.2	549	2.5	870	7.7
6	USGS	21	0.6	9.0	64	1.3	20.3				369	6.6	357	0.3		
7	Army	16	0.5	7.8	53	1.1	19.1	1.2	0.0	0.3	326	5.8	306	0.4	540	7.1
8	Selbach	37	1.0	18.2	120	2.5	43.5				291	5.7	348	0.5	570	7.7
9	17/49	86	2.4	17.6	231	4.8	35.0	28	0.5	3.3	715	13.8	807	−1.4	1250	7.8
10	Rasmus	440	12	27.8	1310	27	61.1				2052	44.6	2851	1.1	3920	7.8
11	IndSpg	4	0.1	2.6	16	0.3	7.8	1.1	0.0	0.4	259	4.4	214	−0.0	400	7.4
12	TW-10	5	0.1	3.8	14	0.3	7.6	1.6	0.0	0.7	221	3.7	186	0.8	350	7.2
13	D. Hole	22	0.6	8.5	78	1.6	22.4	0.3	0.0	<0.1	406	7.3	397	0.1	690	7.8
14	CC Spg	1			9.8											
15	17/52	21			63											

LAB 6

POROSITY, SPECIFIC YIELD, AND SPECIFIC RETENTION

PURPOSE: To gain an appreciation of the concepts of porosity, specific yield, and specific retention through laboratory measurement of these properties.

OBJECTIVES: Measure the porosity of several sand samples.

Determine the specific retention and the specific yield of the same sands.

Investigate the relationship between these properties and the physical characteristics of grain size and sorting.

SAND SAMPLES

Prior to beginning the experiment, examine each sand sample with a hand lens. Estimate and record the size and sorting characteristics of each sample. Recall the clear distinction between *well sorted* (geological terminology) and *well graded* (engineering terminology). Determine average grain diameter (mm) by use of a sand sample card, or align numerous grains along the edge of a mm scale and count the number of grains over several cm.

Description of the samples:

A:

B:

C:

D:

DETERMINATION OF POROSITY

Porosity of a medium can be determined by either of two general methods: volumetric or gravimetric. In the volumetric method, porosity is determined by measuring any two of three quantities: bulk volume, pore volume, and solid volume. In the gravimetric method, porosity is determined by measuring the bulk density of the porous medium and assuming an average grain density. In this lab, you will use both methods.

EQUIPMENT:

Porosimeter

Brass tamping rod

Graduated cylinders, 1000 ml, 250 ml

Sand, 4 different samples

Scale

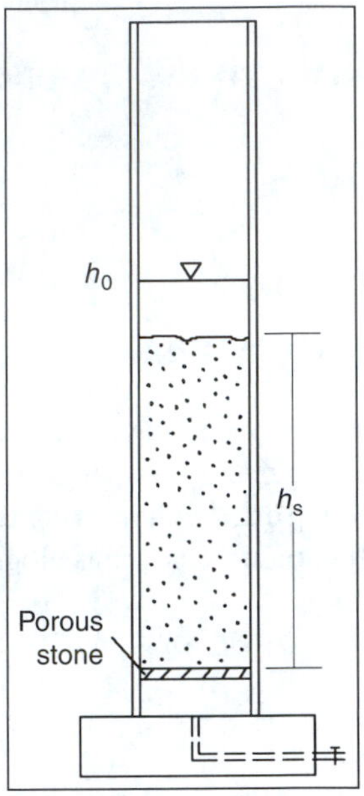

Figure 6.1—Porosimeter.

APPROACH:

You will work with a very simple porosimeter—basically a cylinder capable of holding sand and water (Fig. 6.1). You will define a *control volume*, a known volume of 1000 ml. By adding sand to a known amount of water in this control volume, you can determine the volume of sand, from which you can derive porosity.

PROCEDURE

Define porosity[1] (n)

$n \equiv$

Define the control volume

1. Drop the O-ring into the porosimeter.

2. Close the outlet of the porosimeter, and, using the graduated cylinder, add 1000 ml of water.

3. Remove all air from the porosimeter.

   Tilt the cylinder about 45 degrees toward the drain outlet so air from the drain can escape upward.

   Gently tap the porous stone with the plastic end of the brass rod to free all air bubbles below and around the stone.

   Carefully remove the rod, tapping off any free water into the porosimeter (you should do this each time you use the rod in order to minimize losses).

4. Measure and mark the height of the water (h_0).

5. Drain the porosimeter into the graduated cylinder, and note the volume of water recovered. This amount should

[1]All symbols used in this lab manual are given in Appendix 1.

be within 5 ml of 1000 ml, with the difference due to water adhering to surfaces in the porosimeter and the porous stone. This amount of fluid is equivalent to dead space in the porosimeter.

6. Refill the graduated cylinder with 1000 ml of water, and pour it into the porosimeter.

7. Purge all entrapped air (as in step 3, above).

8. Use a bit of tape or a nonpermanent marker pen to mark this reference water level, h_0. This defines the top of the 1000-ml control volume.

9. Measure the new height of the water, h_0. Is there a difference from the level determined in step 4? Why?

10. Record this reference h_0 in Table 6.1 on p. 49.

Determine the cross-sectional area of the porosimeter

11. Drain about 550 ml of water from the porosimeter into the graduated cylinder.

    Record the change in water height, Δh, and the volume of water drained, ΔV.

    From this, calculate and record the cross-sectional area of the porosimeter.

12. Go to Step 15. (Procedures 13 and 14 will be done only on subsequent runs with different samples; skip these steps on the first sample run.)

13. Top up the graduated cylinder to 1000 ml.

14. Add about 450 ml of water to the porosimeter from the graduated cylinder. Purge all air from the system.

Load sand and water to fill control volume

15. Using the brass rod, seat the O-ring tightly and uniformly around the top of the porous stone.

16. Weigh out a sample of dry sand (about 1250 g). In this next step, you will add dry sand to the cylinder, keeping the level of the sand below the water surface to avoid trapping air. The objective is to *bring the sand up to within 1 cm of the water level and the water level to within 1 cm of* h_0. **Do not raise the water level above h_0.** You probably won't need all the sand.

17. Slowly add sand to the porosimeter. The water level must be maintained above the top of the sand. If needed, add more water from the graduated cylinder. Stop adding sand when the water level is within 1 cm of h_0.

18. Level the surface of the sand, and compact the sand column very slightly using the blunt end of the brass rod.

19. Weigh the unused dry sand remaining. Determine the weight of sand in the porosimeter by difference. Record this weight (W_s) in Table 6.1.

20. Slowly pour water from the graduated cylinder down the inside wall of the porosimeter (to avoid disturbing the sand) until the water level is at the h_0 level. At this point, the volume of water remaining in the graduated cylinder is equal to the volume of the sand. Record V_s.

Calculate porosity by the volumetric method

21. Measure and record the height of the sand in the soil column (h_s).

22. Calculate and record the bulk volume (V_b) of the sand sample (h_s × cross-sectional area − 3.6 ml [volume of O-ring]).

23. Calculate the porosity determined volumetrically (n_v).

Calculate porosity by the gravimetric method

24. Use the dry weight of the sand and the sample volume to determine the (dry) bulk density of the sample (ρ_b).

25. Calculate porosity using the gravimetric method ($n_g = 1 - \rho_b/\rho_s$) assuming a particle density of the sand of 2.65 g/ml. Record the porosity determined gravimetrically (n_g).

DETERMINATION OF SPECIFIC RETENTION AND SPECIFIC YIELD

[Note: Do this procedure only for relatively coarse sands, medium-grained or coarser, depending on sorting. For fine sands and silts or poorly sorted sands, the amount of time required to drain the sample may be hours or days, and you would not see any results during the lab exercise (some lab procedures use a vacuum below the sample to shorten measurement time)].

Define specific retention and specific yield

$S_r \equiv$ $S_y \equiv$

Drain the water from the pores of the sand (and the water in the porosimeter outside the sand)

26. The volume of water now in the sample cylinder is $V_w = 1000 \text{ ml} - V_s$. Record V_w.

27. The volume of water *not* in the pores of the sand is $V_x = 1000 \text{ ml} - V_b$. Record V_x.

28. Drain the porosimeter into an empty 250-ml graduated cylinder.

    Place the graduated cylinder under the outlet.

    Open the outlet valve, slowly but completely.

    Record the volume of water in the graduated cylinder (V_{gc}) each minute while draining.

    Continue monitoring the volume in the cylinder as long as flow remains significant.

Calculate S_r and S_y

29. The volume of water drained from the sand is $V_d = V_{gc} - V_x$.

30. The volume of water retained in the sand is $V_r = V_w - V_{gc}$.

31. Record the final volume of water recovered from the sand.

32. Calculate and record specific retention and specific yield. If any of the samples was still draining when you stopped, plot the values as a function of time and estimate the final value of each parameter by observing an asymptote (note that there may be some nonsense values caused by the lab setup and procedure; do not plot nonsense values).

Name ____________________ Date ____________________

Table 6.1—EXPERIMENTAL DATA					
Sample ID					
h_0 (cm)					
$\Delta V/\Delta h$ (cm^3/cm)					
Area, cross section (cm^2)					
W_s (g)					
V_s (cm^3)					
h_s (cm)					
$V_{bulk} = h_s \times$ area (cm^3)					
$n_v = (V_{bulk} - V_s)/V_{bulk}$					
$\rho_b = W_s/V_{bulk}$					
$n_g = 1 - \rho_b/\rho_s$					
$V_w = 1000\text{ cm}^3 - V_s$					
$V_x = 1000\text{ cm}^3 - V_b$					
1 min	V_{gc} (cm^3)				
	$V_d = V_{gc} - V_x$				
	$V_r = V_w - V_{gc}$				
2 min	V_{gc} (cm^3)				
	$V_d = V_{gc} - V_x$				
	$V_r = V_w - V_{gc}$				
3 min	V_{gc} (cm^3)				
	$V_d = V_{gc} - V_x$				
	$V_r = V_w - V_{gc}$				
4 min	V_{gc} (cm^3)				
	$V_d = V_{gc} - V_x$				
	$V_r = V_w - V_{gc}$				
5 min	V_{gc} (cm^3)				
	$V_d = V_{gc} - V_x$				
	$V_r = V_w - V_{gc}$				
End	V_{gc} (cm^3)				
	$V_d = V_{gc} - V_x$				
	$V_r = V_w - V_{gc}$				
	S_r				
	S_y				

33. Dump the porosimeter and rinse it out preparatory to measuring the next sample. When removing the wet sand, do *not* return it to the original (dry) container; place it in the labeled wet-sand container. AVOID MIXING SAMPLES.

Repeat these procedures for each of the lab samples (go back to step 13).

---

34. Exchange quantitative results with others in the lab, compiling data in this table, and calculate mean (μ) and median (M) values and standard deviation (σ):

		A	B	C	D
Porosity	Values				
	μ, M, σ				
S_r	Values				
	μ, M, σ				
S_y	Values				
	μ, M, σ				

LAB REPORT

Write a brief report on the *results* of this lab. Ensure that you report significant figures only.

SUMMARY				
Sample	**A**	**B**	**C**	**D**
Size				
Sorting				
Porosity				
Specific retention				
Specific yield				

1. Correlate porosity, specific retention, and specific yield with grain size and sorting.
2. Examine possible sources of error in measurement and the relative importance of each. Suggest improvements in techniques.
3. Compare porosities obtained by the gravimetric and volumetric methods. Which method is better, and why?
4. Explain how, in terms of experimental setup, a specific yield result might be negative.
5. Is the volumetric method used here suitable for determination of porosity, specific retention, and specific yield in naturally occurring rocks and soils?

LAB 7

DARCY'S LAW AND HYDRAULIC CONDUCTIVITY

PURPOSE: Gain an understanding of the controls on groundwater flow through porous media, using a simulation of Darcy's experiment.

OBJECTIVES: Determine the hydraulic conductivity and permeability of four sand samples.

Determine the relationship between discharge and hydraulic gradient.

Determine the relationship between discharge and hydraulic conductivity.

Determine the relationship between permeability and hydraulic conductivity.

Relate the magnitude of hydraulic conductivity to grain size and degree of sorting.

Become familiar with laboratory methods for measuring permeability, using both constant–head and falling–head permeameters.

REPORT REQUIREMENTS: A table of observational and computational data.

A summary of hydraulic conductivity, K (15.5°C), and permeability, k.

The relationship between discharge and hydraulic gradient (constant-head permeameter observations).

The relationship between discharge and hydraulic conductivity.

The relationship of hydraulic conductivity to grain size and sorting.

A note concerning any apparent departures from Darcy's law.

READING ASSIGNMENT: In addition to reading through this lab, read material on permeameters in your textbook (e.g., Fetter, 2001, pp. 90–93) and do Problem 3 (on the CD). Summarize your results from Problem 3 in Procedure 14, which follows.

PROCEDURES

Darcy's Law and Hydraulic Conductivity Using Constant-Head Permeameters

You will use these procedures to simulate Darcy's experiments, measuring flow through two different sands under several different heads. If discharge is a linear function of the hydraulic gradient, $\Delta h/\Delta l$, the constant of proportionality is hydraulic conductivity. By measuring flow as a function of head and graphing the results, the slope of a regression line through the origin will be hydraulic conductivity.

1. Characterize the first of your sand samples as to grain size and sorting; record this and other observations on the data sheet provided, Table 7.1 on page 53. Include a measurement of median grain size, which you will use as an approximation of median pore size.

2. The permeameters have been set up already to ensure that the samples do not contain entrapped air. **Do not allow the samples to become unsaturated**, or you will have to spend an additional 15 minutes putting the sample under vacuum to remove the air (and you will be jeered loudly by the rest of the students who will also get out of lab 15 minutes late).

3. Refer to Figure 7.1, which shows the basic arrangement for a constant-head permeameter. Permeameter designs vary somewhat, so your apparatus may look slightly different, but the operation will be the same.

   Measure and record the length of the sand column (Δl).

   Record the cross-sectional area of the sample (A), written on the permeameter.

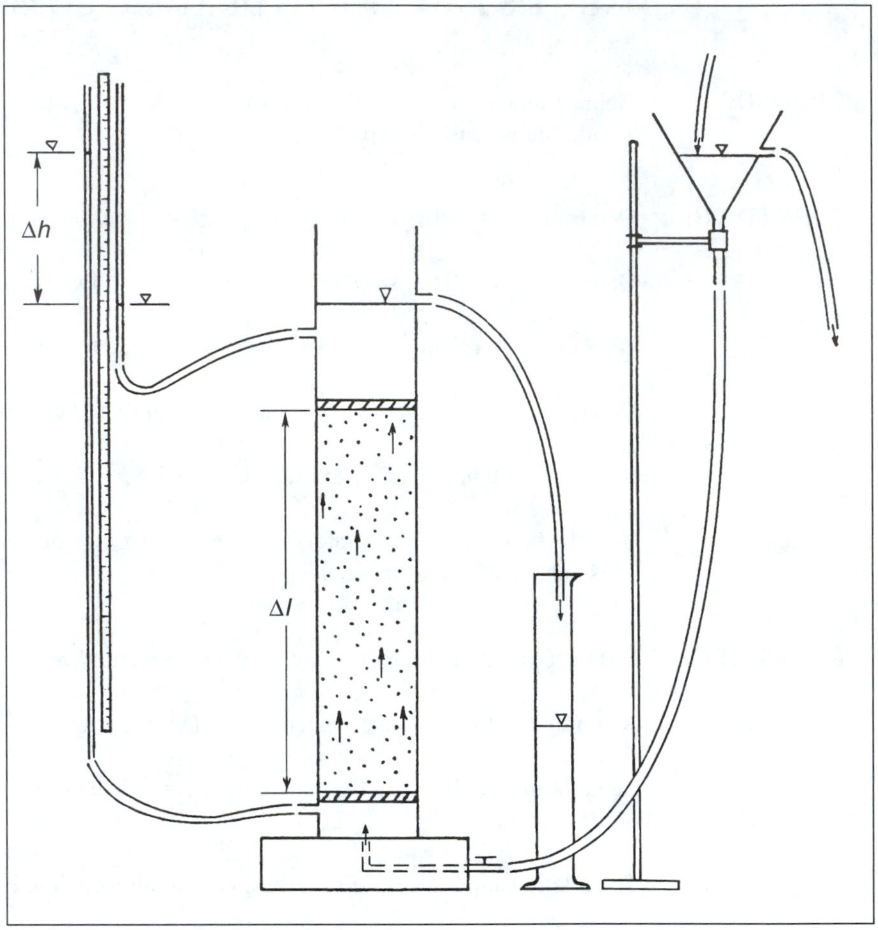

Figure 7.1—Constant-head permeameter.

4. Adjust the height of the funnel so that its overflow port is about 2 cm above the outflow port of the permeameter. This will establish a head difference of about 2 cm.

5. Start a flow of water into the funnel.

6. Start the flow of water through the sand.

   Adjust the inflow to maintain a constant level in the funnel.

7. When flow has stabilized, collect outflow from the permeameter in a graduated cylinder for 3 to 5 minutes, while timing.

   Measure and record the difference in head across the sand column (Δh).

8. Record the volume of water collected (V), the time interval (t), and the temperature (T).

   Calculate and record average discharge (Q).

Table 7.1—CONSTANT-HEAD PERMEAMETER DATA										
	Sample ________					**Sample** ________				
	Length (Δl) _____ cm Area (A) _____ cm^2 Temp (T) ________					Length (Δl) _____ cm Area (A) _____ cm^2 Temp (T) ________				
Run	Δh cm	V ml	t s	Q ml/s	$A\Delta h/\Delta l$ cm^2	Δh cm	V ml	t s	Q ml/s	$A\Delta h/\Delta l$ cm^2
1										
2										
3										
K_T										
$K_{15.5}$	Calculation:					Calculation:				
k	Calculation:					Calculation:				
Sand description:						Sand description:				

9. Repeat steps 4 through 8 twice more, with different head differences, somewhere around 4 cm and 8 cm Δh. High heads may generate high flow-velocities that lead to nonlinear laminar flow or even turbulent flow.

10. Repeat procedures 1 through 9 for the second sample.

The remaining steps of this first procedure can be done after lab, although it is preferable to do them before leaving, in case there were procedural errors.

11. Write Darcy's equation.

12. Graph the results of your lab observations in Figure 7.2.

    Does Darcy's equation appear to be valid? That is, for a given sand, is the rate of flow of water directly proportional to $\Delta h/\Delta l$?

    If so, the slope of a regression line through the origin is the constant of proportionality, K, or hydraulic conductivity. Actually, it is K_T, the hydraulic conductivity at temperature T.

    Record K_T for each sample.

13. Hydraulic conductivity should be reported at a standard temperature, taken to be 15.5°C. The fluid properties that are temperature-sensitive are density and viscosity. By looking at the values in Table 7.2 on page 55, you can see that density variations are very small compared to changes in viscosity, and, in fact, density can be ignored.

    Thus, K_T can be corrected to K (15.5°C) by

$$\frac{K_T}{K_{15.5°C}} = \frac{\mu_{15.5°C}}{\mu_T}$$

    Calculate and record $K_{15.5}$.

14. Permeability, k, can now be derived from K. The conversion from K to k is easier if you use the conversions you determined in Problem 3.

    Summary: At 15.5°C, 1 μm^2 = darcies = m/day = gal/day/ft^2
    Calculate and record permeability.

15. For each of the tests, calculate a Reynolds number from $N_R = \rho V d/\mu$, where V is specific discharge (not velocity) and the median grain diameter, d, is taken to be the median pore diameter.

    Does each test appear to be valid in terms of probable linear laminar flow—that is, is the flow Darcian? Record results in Table 7.4, on page 58.

16. Estimate the sources of error in this procedure.

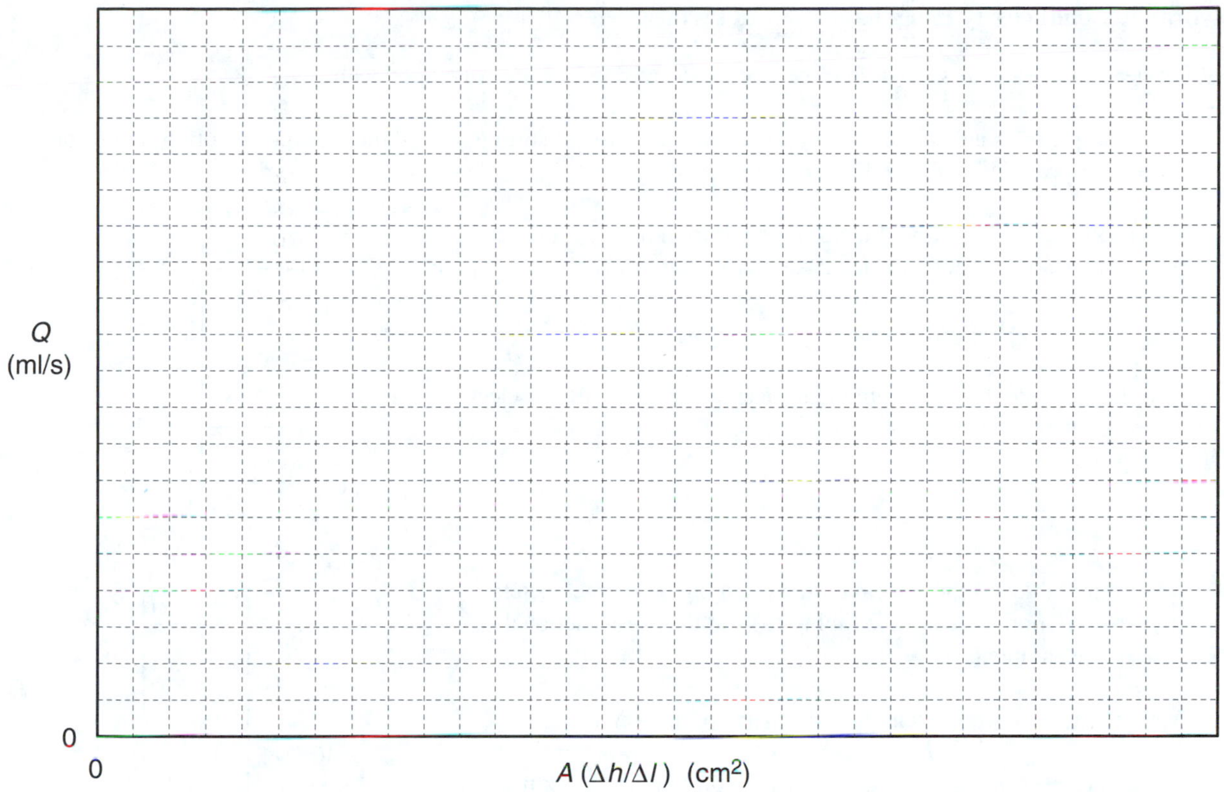

Figure 7.2—Relationship of discharge to hydraulic gradient.

Table 7.2—DENSITY AND VISCOSITY OF WATER						
Temperature (°C)	**Density (g/ml)**	**Viscosity (millipoise)**		**Temperature (°C)**	**Density (g/ml)**	**Viscosity (millipoise)**
0	.9999	17.94		20	.9982	10.09
1	.9999	17.32		21	.9980	9.84
2	1.0000	16.74		22	.9978	9.61
3	1.0000	16.19		23	.9976	9.38
4	1.0000	15.68		24	.9973	9.16
5	1.0000	15.19		25	.9971	8.95
6	1.0000	14.73		26	.9968	8.75
7	.9999	14.29		27	.9965	8.55
8	.9999	13.87		28	.9963	8.36
9	.9998	13.48		29	.9960	8.18
10	.9997	13.10		30	.9957	8.00
11	.9996	12.74		31	.9954	7.83
12	.9995	12.39		32	.9951	7.67
13	.9994	12.06		33	.9947	7.51
14	.9993	11.75		34	.9944	7.36
15	.9991	11.45		35	.9941	7.21
15.5	**.9990**	**11.30**		36	.9937	7.06
16	.9989	11.16		37	.9934	6.92
17	.9988	10.88		38	.9930	6.79
18	.9986	10.60		39	.9926	6.66
19	.9984	10.34		40	.9922	6.56

Note: 1 poise = 1 dyne sec cm^{-2} = 1 g cm^{-1} sec^{-1} 1 poise = 2.09 E-3 lb sec ft^{-2}

Hydraulic Conductivity Using Falling-Head Permeameters

Falling-head permeameters are used mostly for fine-grain sediments of low permeability.

1. Characterize the sand sample as to grain size and degree of sorting; record these and other observations on the data sheet provided, Table 7.3 on page 57.

2. Refer to Figure 7.3. You will measure the drop in head over time while the head continuously decreases across the sand column.

   In order to make these measurements accurately, you will preselect reference marks near the top and bottom of the tubing to use as timing marks.

   Record in Table 7.3 the information written on the permeameter: the cross-sectional area of the tubing (a) and the cross-sectional area of the sample (A).

3. Measure and record the length of the sand column (Δl).

4. Select a reference point near the top of the tubing, and mark it with a nonpermanent marker pen or a bit of tape. Measure and record the distance from this mark to the bottom of the sand column[1] (h_0).

5. Select a reference mark near the bottom of the tubing a few centimeters above the top of the permeameter, and mark it also.

   Measure and record the distance from this reference point to the bottom of the sand column h_1.

6. Calculate and record $(h_0 h_1)^{1/2}$, and mark this point on the tubing.

7. Fill the tubing to the top.

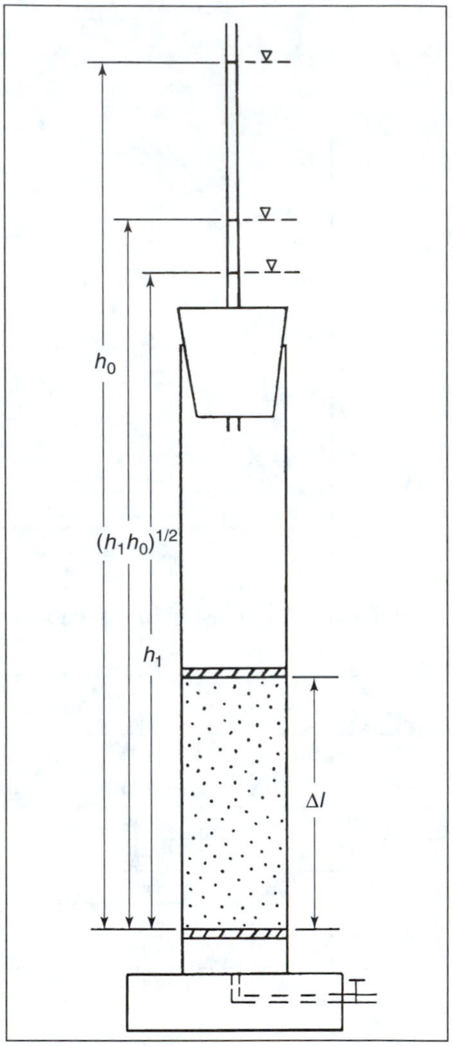

Figure 7.3—Falling-head permeameter.

[1]With low-permeability samples for which this apparatus is designed, the space below the sand column is normally drained almost immediately. If, however, this is *not* the case, the reference point for head measurements should be the discharge point.

Table 7.3—FALLING-HEAD PERMEAMETER DATA

Sample ________	Tubing Area (a): ________ Sample Length (Δl): ________ Sample Area (A): ________ Temperature (T) ________					
Run	Start				End	
	h_0 ________		$(h_0h_1)^{1/2}$ ________		h_1 ________	
	Time h_0	Δt $h_0 - (h_0h_1)^{1/2}$	Time $(h_0h_1)^{1/2}$	Δt $(h_0h_1)^{1/2} - h_1$	Time h_1	Δt $h_0 - h_1$
1						
2						
3						
Average						
K_T						
$K_{15.5}$						
k						
Sand description						

Sample ________	Tubing Area (a): ________ Sample Length (Δl): ________ Sample Area (A): ________ Temperature (T) ________					
Run	Start				End	
	h_0 ________		$(h_0h_1)^{1/2}$ ________		h_1 ________	
	Time h_0	Δt $h_0 - (h_0h_1)^{1/2}$	Time $(h_0h_1)^{1/2}$	Δt $(h_0h_1)^{1/2} - h_1$	Time h_1	Δt $h_0 - h_1$
1						
2						
3						
Average						
K_T						
$K_{15.5}$						
k						
Sand description						

In the next step, you will measure the time required for the head to drop from h_0 *to* h_1*, with an intermediate reading at* $(h_0h_1)^{1/2}$*. Read through steps 8 through 11 before starting.*

8. When ready to begin timing, open the clamp to start water flow. Start timing when the water level reaches the top reference mark (h_0).

9. Observe and record the time for the water level to fall to the $(h_0h_1)^{1/2}$ point.

10. When the water level in the tubing falls to the bottom reference point (h_1), **stop timing and close the clamp! Do not allow the sand column to become unsaturated** (see note under step 2 of constant-head procedure).

11. Record the times and the temperature (T).

12. The time required for head decline from h_0 to $(h_0h_1)^{1/2}$ should about equal the time for head decline from $(h_0h_1)^{1/2}$ to h_1 (agreement decreases as clay content of the sample increases).

13. Repeat steps 7 through 12 to obtain three sets of data.

---

14. Switch to the other permeameter and repeat steps 1 through 13 for the second sample.

---

15. K can be calculated from the falling-head permeameter equation

$$K_T = \frac{a}{A}\frac{\Delta l}{t}\ln\frac{h_0}{h_1}$$

16. Correct K for temperature, and calculate permeability.

17. For each of the tests, calculate a Reynolds number, using $N_R = \rho V d/\mu$, where V is specific discharge (not velocity) and d, the median grain diameter, is taken to be the median pore diameter.

    Does each test appear to be valid in terms of probable linear laminar flow—that is, is the flow Darcian? Record answers in Table 7.4.

18. Estimate sources of error in this procedure.

19. Complete Table 7.4.

Table 7.4—SUMMARY

Sample ID	Grain Size	Sorting	K (m/day)	k(μm^2)	N_R	Darcian?

20. Describe the relationship of permeability to grain size.

21. Describe the relationship of permeability to sorting.

22. Darcy apparently did not recognize the effect of fluid properties on flow. Does the temperature of the water affect hydraulic conductivity significantly? If so, what temperature-dependent property of water is most influential?

LAB 8

MODELING GROUNDWATER FLOW WITH FLOWNETS

PURPOSE: Become familiar with a graphical method for modeling flow through an aquifer with irregular boundaries. In one exercise, the boundaries are well defined, whereas in the second exercise you will need to define the boundaries of the flow field using subsurface geologic techniques.

OBJECTIVES: Construct a vertical cross-section flownet to determine groundwater flow under a dam and to calculate uplift pressure on the dam.

Construct a horizontal (map view) flownet to determine regional flow through a deep, confined aquifer.

PROCEDURE: Before beginning either of the flownet problems, read through the material below on flownet construction, with guidelines and suggestions for making a flownet.

FLOWNET CONSTRUCTION

Although the following procedures are more or less sequential, be certain you read through and understand *all* the procedures and suggestions before you begin to actually construct a flownet.

1. Ensure that the use a flownet is valid.
   (a) Flow is steady state.
   (b) The third dimension is constant.
   (c) Boundary conditions are known or can be approximated well.

2. Prepare for a trial-and-error solution; there is no *unique* solution.

3. Ensure that the boundaries of the flow field are drawn to scale.

4. Define the boundaries, and determine the nature of each boundary.
   (a) equipotential, or constant-head, boundary
   (b) flow line, or no-flow, boundary; commonly an impermeable boundary
   (c) known-head boundary; commonly the water table
   (d) most often encountered—two equipotential boundaries and two flow line boundaries.

5. Determine the symmetry of the flow field, if any. This may reduce the work involved—for example, bilateral symmetry would require construction of only half of the flownet.

6. Begin iterative construction of flow lines and perpendicular equipotential lines.
   (a) Draw flow directions at equipotential boundaries.
   (b) Draw predefined, or boundary, flow lines.
   (c) Visualize the infinite number of flow lines that provide a smooth transition between the boundary flow lines.
   (d) You will draw only a few of these, perhaps two or three.

7. A fundamental requirement of a flownet is that flow in each of the flow tubes must be equal. Thus, because tubes that are longer will have a lower hydraulic gradient, they must therefore be wider in order to carry the same flow.

   For this reason, when you draw a flownet that curves (e.g., the flow field under the dam in Figure 8.1 on page 62), you must draw flow tubes that are progressively wider with an increase in path length or radius of curvature. In Figure 8.1, flow tubes will widen with depth.

8. In terms of flownet construction, a simpler restatement of the above is **in order to maintain "squares" ($\Delta w = \Delta l$), Δw must increase as Δl increases**.

9. Start at the most constricted part of the flow field, and lightly sketch a few flow lines and equipotential lines.

10. **Each area you construct must be correct before continuing** (if there is any error, all further work will be pointless).

11. Construct smooth lines that define equidimensional enclosures, or "squares."

    You can check this by drawing circles inside the enclosures that are tangential at the four midpoints, or you can connect midpoints of the sides with curved lines and see if they are of equal length.

12. All intersections of lines must be orthogonal.

13. Adjust by trial and error. Your first attempt at drawing a flownet like that in Figure 8.1 may take an hour or more.

14. Maintain a sense of humor.

15. If the number of flow tubes chosen is an integer, it will be only by coincidence that the number of equipotential drops will be an integer. A corollary to this is that you should expect to finish the flow field at an equipotential boundary with a fraction of a square.

16. When your flownet looks complete, you can check your construction by lightly sketching diagonals across the squares. These diagonals also will form a flownet that is smooth, encloses squares, and has perpendicular intersections.

You can do it this way, or you can do it your way. There are no required steps or procedures, but the suggestions above should help you avoid problems. A net that goes astray can usually be traced back to a violation of one of the above steps (most commonly to steps 8 and 10).

In each of the exercises that follow, before starting any flownet construction, *think the problem through.*

GROUNDWATER FLOW BENEATH A DAM

Figure 8.1 shows a vertical cross section through a typical concrete gravity dam with a cutoff trench. You are asked to determine the flow of water under the dam and to calculate uplift pressure at the bottom of the dam. The solution to both problems will require that you construct a flownet to simulate flow beneath this dam.

1. Before beginning actual construction, work for a minute with the smaller figure to the right.
   (a) What are the boundaries to this flow field? Label each of the four boundaries, noting the type of boundary.
   (b) Draw the two flow lines that constitute the flow line boundaries, showing direction of flow.
   (c) Visualize numerous flow lines between them, all of which form a smooth transition from a simple, straight flow line at the bottom to a very angular flow line at the top. Sketch several of these flow lines, enough to show flow directions throughout the aquifer.

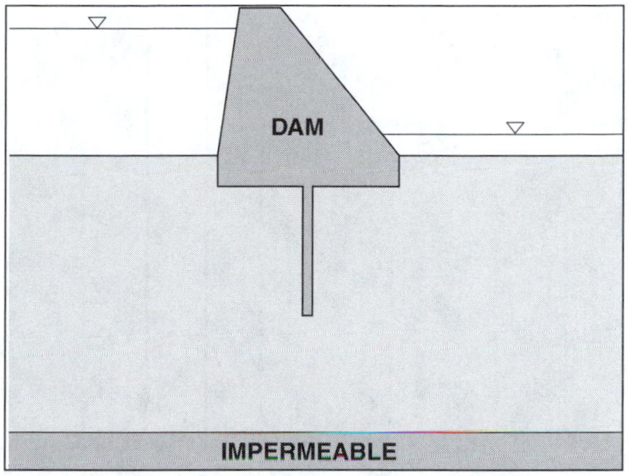

2. Construct a flownet on Figure 8.1 showing groundwater flow under the dam.

3. The dam in Figure 8.1 is 100 m long (along the axis), and the abutments are in solid bedrock. Aquifer tests have shown the alluvium has a hydraulic conductivity of 0.34 m/day. The reservoir depth is maintained constantly at 12 m, with a tailwater pool 2 m deep.

   How much water is lost under the dam (Q)?

   This can be calculated by

$$Q = \frac{n_s}{n_d} KwH$$

   where Q is discharge
   n_s is the number of stream tubes, or flow tubes, in the net
   n_d is the number of potential drops across the net
   K is hydraulic conductivity
   w is the width of the aquifer (thickness in the third dimension, perpendicular to the page)
   H is the total head loss across the flow field

4. Gravity dams depend on their mass for stability. Gravity dams can fail by sliding or by tipping over—that is, by rotation about the toe if lateral pressures on the upstream face become excessive. The probability of either type of failure is increased by uplift pressures on the bottom of the dam, especially near the upstream face. In order to design a safe dam, therefore, uplift pressures must be known.

   What is the uplift pressure at point A?

   Solution: Determine the pressure head at A, from which, knowing the specific weight of water, you can calculate pressure. Pressure head, you'll recall, is total head minus elevation head. You can simplify the calculation by using the bottom of the dam as your datum, so elevation head is zero. The head at A can be read from your constructed flownet.

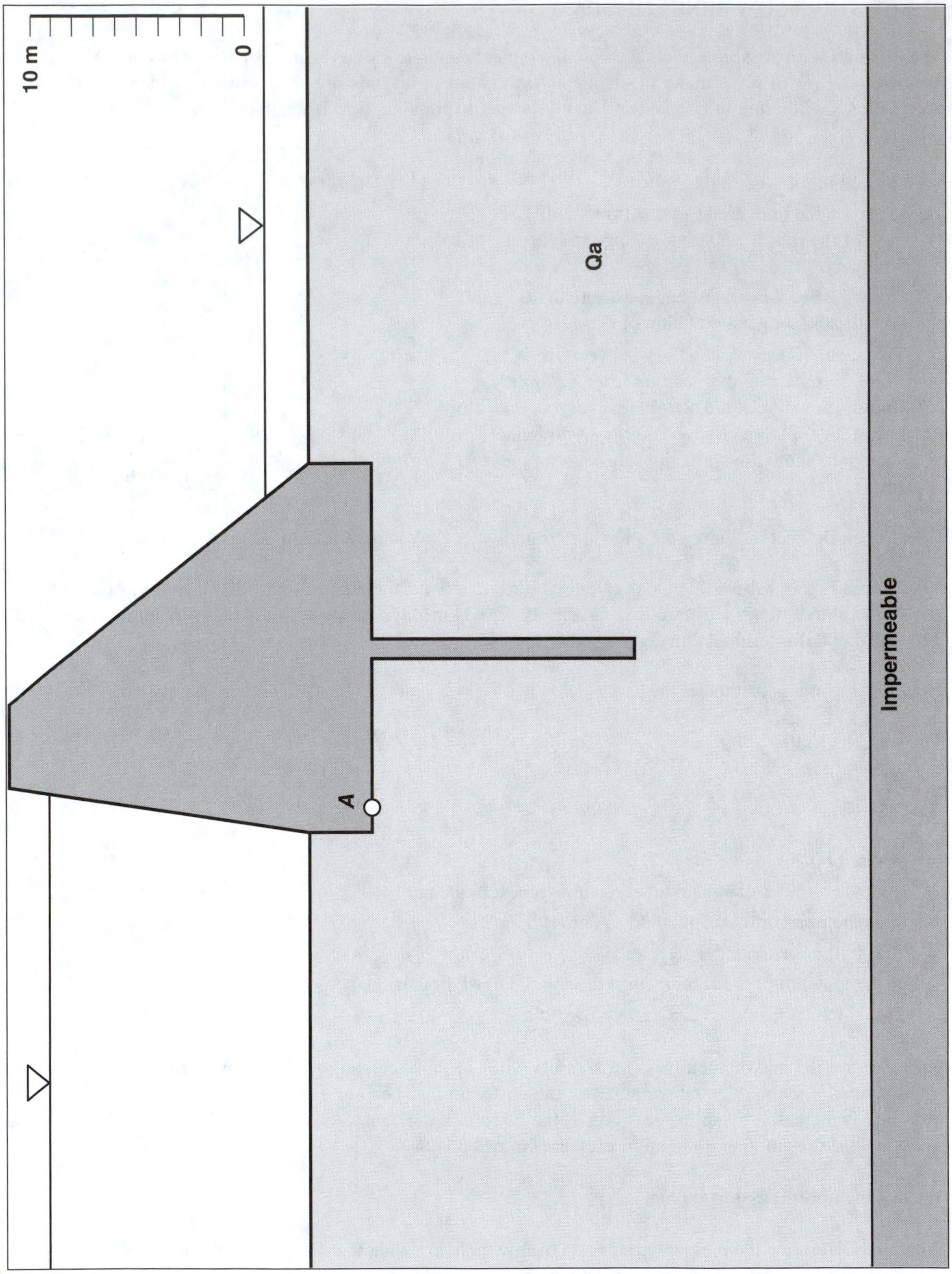

Figure 8.1—Cross section of gravity dam on alluvium.

GROUNDWATER FLOW THROUGH A DEEP, CONFINED AQUIFER

Subsurface information is available from logs of three wells in this area, as shown in Figure 8.2. The well data are presented in a format common for such data—all measurements are made relative to the ground surface—that is, they are *depth* measurements.

In order to correlate between wells, you will need to reduce the depth/ground elevation data to a common datum, normally sea level, because that is the datum for ground elevations.

Each of the wells was cased, with the casing perforated in the intervals shown. The map in Figure 8.3 shows structure contours on the top of impermeable bedrock. This is all of the information available.

1. Determine the discharge through the aquifer.

Suggestions: Note in looking at Figure 8.3 that there are no boundaries to a flow field shown, nor do you see any equipotential surfaces. One main part of this problem, therefore, will be to determine the aquifer geometry. You can do this by working with the well data to determine the aquifer depth and thickness, and solution of a three-point problem will provide an approximate strike and dip of the aquifer. You will have to combine this information with the structure contour map to determine the aquifer geometry and, thus, the boundaries of the flow field.

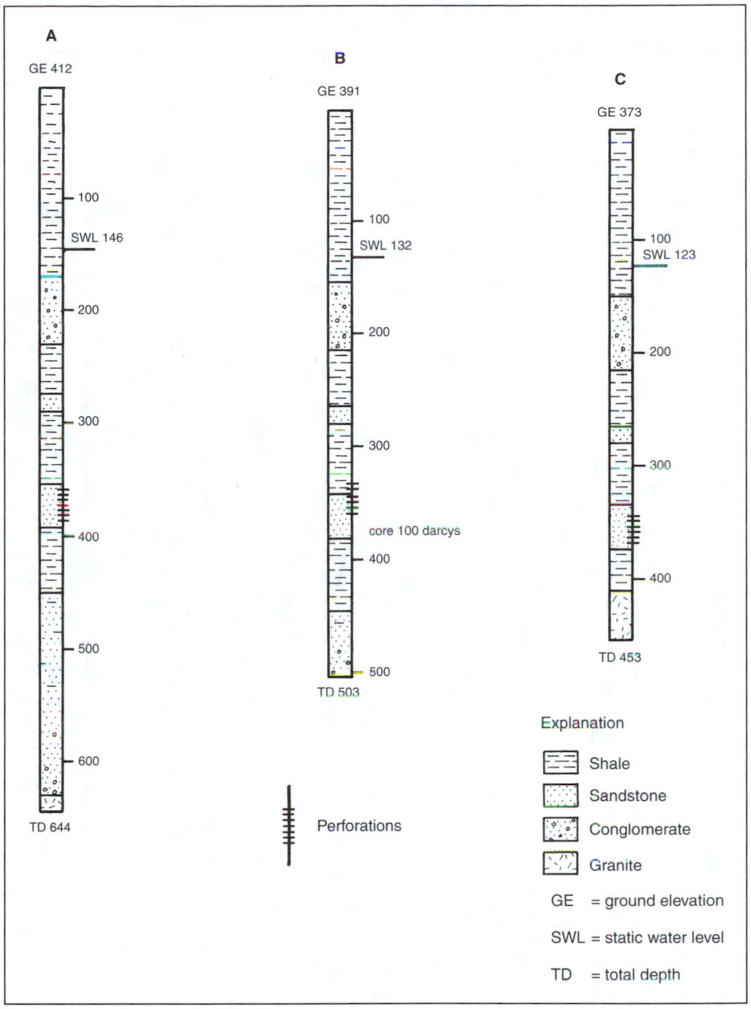

Figure 8.2—Lithologic logs of the three wells shown on the map (Fig. 8.3) plotted to scale. Measurements are in feet; there is no horizontal scale.

Try to visualize the aquifer in three dimensions. This is difficult, but start with the dip. Which way is the aquifer dipping? You might begin by correlating the lithologic units in the wells. Draw lines connecting the tops of each unit, including the basement.

The general dip direction of the sedimentary rocks is toward the __________.

You can get a more quantitative estimate of strike and dip by solving a three-point problem, as shown below.

Given three points of known elevation on a plane at *A*, *B*, and *C*, you can determine the strike by connecting any two points of equal elevation. For example, draw a line connecting the two points of highest and lowest elevation, in this example *A* and *C*, and subdivide it into equal increments. A line connecting *B*, at 200 ft elevation, with the point on the line at 200 ft, defines the strike—here N 70 W.

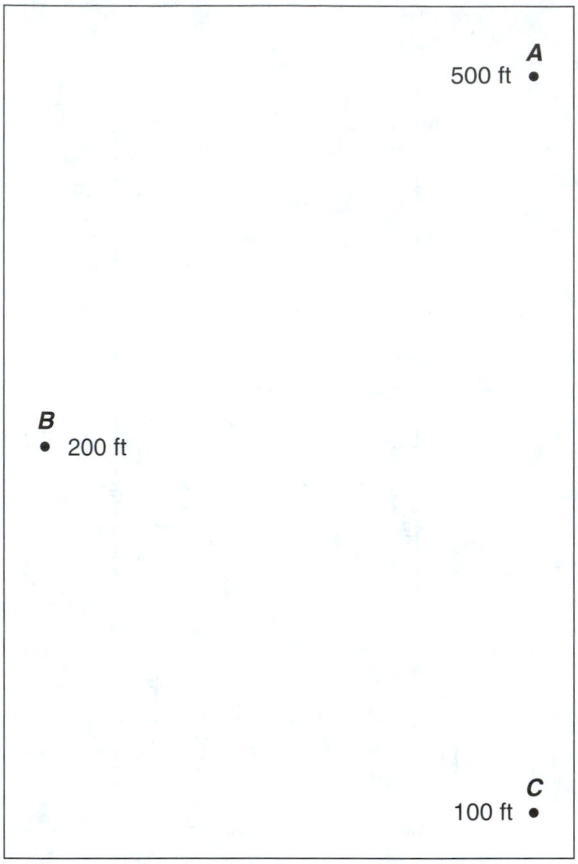

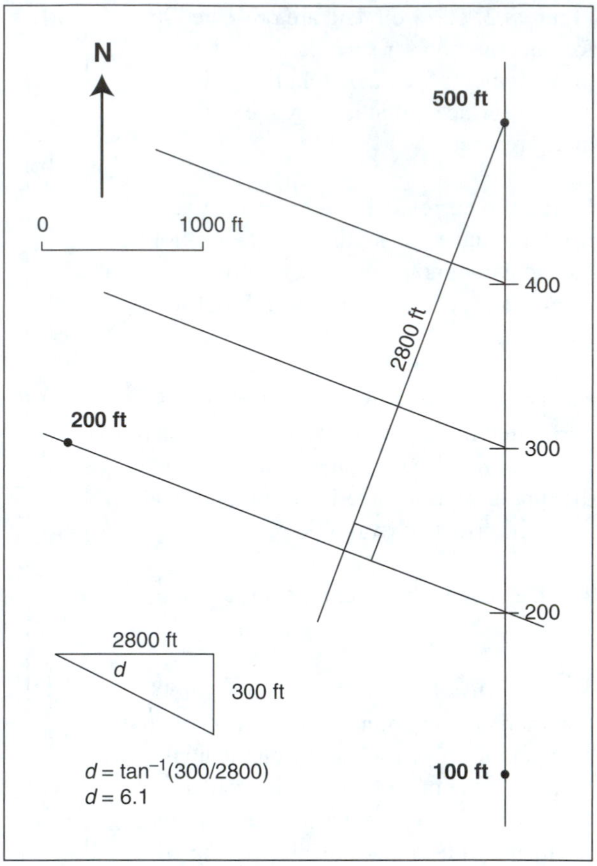

Dip can be determined trigonometrically: Construct a line from *A* perpendicular to the strike line, and measure its length—here 2800 ft. Over this distance, the plane drops 300 ft, so the dip is calculated as shown.

The plane has a strike and dip of N 70 W, 6.1° SW.

Now, returning to the main problem, the strike of the aquifer is __________ and the dip is __________.

Will the aquifer be a continuous sheetlike body throughout the area, or was its deposition limited by paleotopography? Your correlation lines on Figure 8.2 will suggest an answer.

The elevation of the top of the aquifer in Well *C* is __________.

How does this relate to the paleotopography?

You should be able now to approximate the extent of the sandstone aquifer. You could get a closer approximation by superposing a structure contour map of the aquifer onto the structure contour map of the granite (note that your three-point problem solution *is* a structure contour map).

Although potentiometric surfaces rarely are truly planar, you can get a first approximation of the surface by assuming it is planar. As a three-point problem solution estimates the strike and dip of a bed, potentiometric data from three wells can be used to estimate the strike and hydraulic gradient of the potentiometric surface.

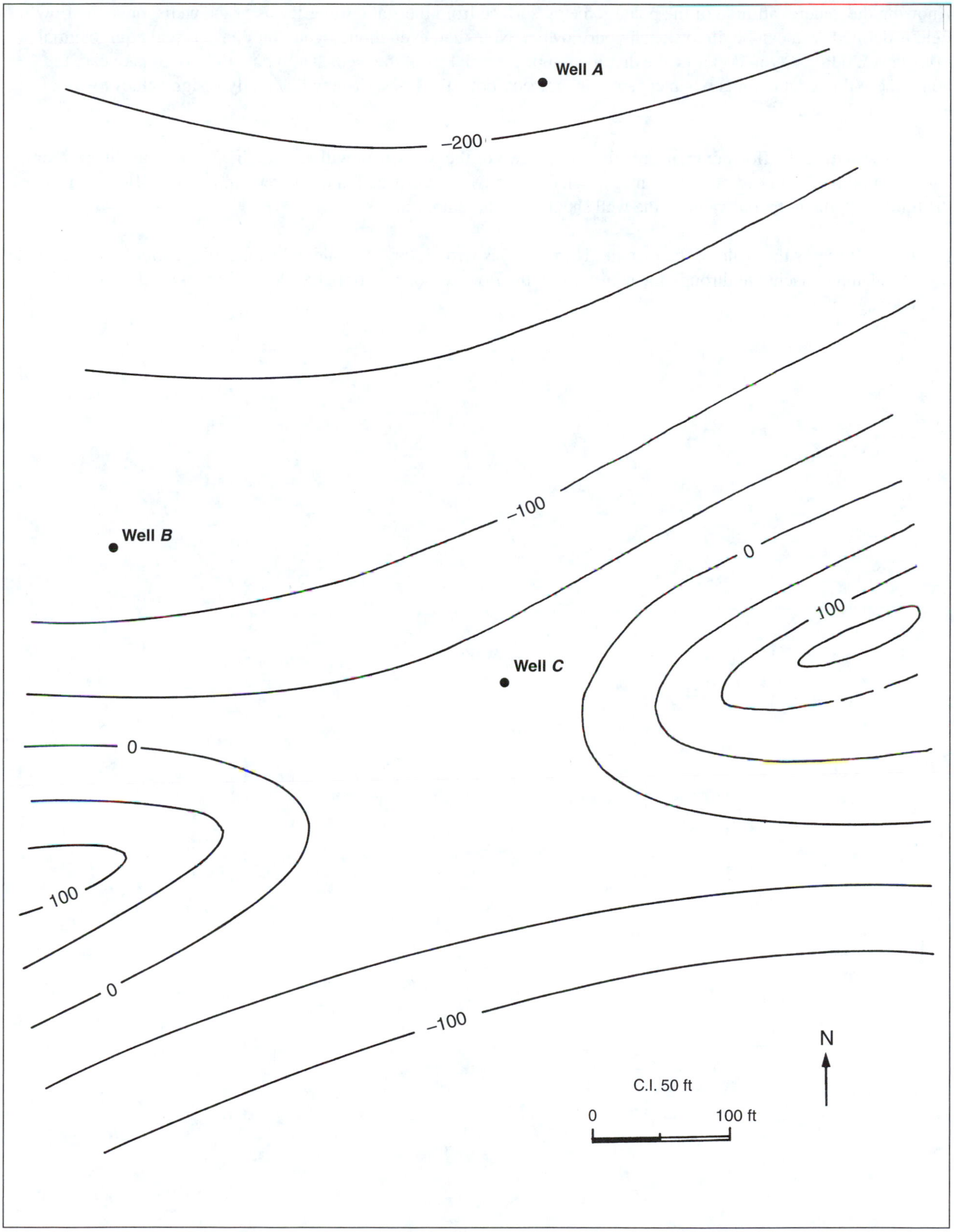

Figure 8.3—Structure contour map of the top of Precambrian granite (contour interval 50 feet).

Knowing the general attitude of the potentiometric surface from the static water levels of the wells, once the flow field is defined, you can begin to sketch your flownet. Note that, even though you don't have actual equipotential *boundaries* defined, you do know the distribution of potential from the wells, and you will have to take care to honor these three data points when constructing the equipotential lines. (You will actually assign values to your equipotential lines.)

After constructing the flownet in an area that covers two of the wells, you will have defined a contour interval for the equipotential lines (it need not be an integer value). When your construction of the remainder of the flownet reaches the third well, the potential value at the well should match your contours.

2. Oops! It seems the scale of the map in Figure 8.3 was wrong; the bar scale indicating 100 ft should read *200 ft*. Recalculate discharge through the aquifer. Will the flow be doubled or halved? Or will you need to start all over again?

LAB 9

AQUIFER TESTING I: METHODS FOR ANALYZING AQUIFER TEST DATA

OVERVIEW OF AQUIFER-TESTING LABS: Labs 9–12 deal with methods of testing aquifers to determine their hydraulic properties, primarily hydraulic conductivity or transmissivity and storativity. Lab 9 deals with methods designed for ideal, infinite-extent, confined aquifers. Lab 10 treats departures from these conditions, specifically semiconfined and unconfined aquifers and aquifers of limited extent. Lab 11 deals with quick and inexpensive slug tests, and Lab 12 is a template for field observations of an aquifer test.

PURPOSE: Learn the Theis and Jacob methods of determining aquifer constants—transmissivity and storativity—from aquifer test (pump test) data.

OBJECTIVES: Determine the aquifer characteristics—transmissivity and storativity—of several aquifers.
Analyze drawdown as a function of time, using both the Theis and the Jacob methods.
Determine the limits and applicability of each method.
Analyze drawdown data from several observation wells at one time, using the Jacob method.
Recognize drawdown data from both confined and unconfined aquifers.
Use these methods in reverse to predict aquifer response to pumping.

MATERIALS: You will need a plot of the "Theis-type curve" ($W[u]$ versus u) on log-log paper and blank log-log paper of the same scale to plot test data. Before lab, access the file *W*(*u*).xls on the CD, which contains a plot of the type curve [Sheet Theis Curve] and a blank log-log plot of s versus t/r^2 [Sheet Data Plot], and print out the type curve and several blank data plots (alternatively, you can use the Theis-type curve in Figure 9.1 on page 69, and overlay tracing paper on it to plot test data). Bring to lab at least one sheet of 4-cycle semilog paper. You will also need a straightedge, colored pencils, and a calculator.

READING ASSIGNMENT: Read through this lab and review the background for these tests by reading your textbook material on aquifer testing using the Theis and Jacob methods (e.g., Fetter, 2001, pp. 169–177).

REVIEW OF THEIS'S SOLUTION OF THE NONSTEADY, RADIAL, CONFINED-FLOW EQUATION

General equation for nonsteady, radial, confined flow (in polar coordinates)[1]

$$\frac{\partial^2 h}{\partial r^2} + \frac{1}{r}\frac{\partial h}{\partial r} = \frac{S}{T}\frac{\partial h}{\partial t}$$

C.V. Theis (1935), by analogy to heat flow, provided a solution for this equation using these assumptions.

1.	Q is constant	(Q can be controlled)
2.	homogeneous, isotropic aquifer	(can handle layered heterogeneity)
3.	aquifer of infinite extent	(useful for extensive aquifers)
4.	well fully penetrating aquifer	(can modify for partial penetration)
5.	infinitesimal well diameter	(finite diameter okay if time is greater than a few minutes)
6.	instant discharge with drop in head	(okay for confined aquifer)

[1]See Appendix 1 for explanation of all symbols.

Although all these assumptions often are not justified in the field, Theis's equation, with modification, has been applied successfully.

Theis equation

$$s = \frac{Q}{4\pi T}\int_u^\infty \frac{e^{-u}}{u}\,du$$

where $u = \frac{r^2 S}{4Tt}$

Note: $s \propto \frac{Q}{T}, \frac{Tt}{r^2 S}$

Note: s, drawdown, is sometimes denoted as $h_0 - h$ (e.g., Fetter, 2001).

This cannot be integrated directly, but the exponential integral

$$\int_u^\infty \frac{e^{-u}}{u}\,du,$$

also known as $W(u)$ (read *well function of* u), is given by the series

$$W(u) = \left[-0.577261 - \ln u + u - \frac{u^2}{2 \times 2!} + \frac{u^3}{3 \times 3!}\cdots\right]$$

Note: $W(u) \propto 1/u$

Graph of $W(u)$ versus $(1/u)$ is called the Theis-type curve.

Values of $W(u)$ versus $(1/u)$ are plotted in Figure 9.1 and are in the CD file $W(u)$.xls.

REVIEW OF THEIS'S GRAPHICAL METHOD FOR DETERMINING FORMATION CONSTANTS *S* AND *T* FROM AQUIFER TEST DATA

From the Theis equation

$$s = \frac{Q}{4\pi T}W(u)$$

where

$$u = \frac{r^2 S}{4Tt}$$

$$T = \frac{Q}{4\pi s}W(u)$$

and

$$S = \frac{4Tu}{r^2/t};\ \frac{t}{r^2} = \frac{S}{4Tu}$$

From the first equation above

$$\log s = \left[\log\left(\frac{Q}{4\pi T}\right)\right] + \log W(u)$$

From the last equation above

$$\log t/r^2 = \left[\log\left(\frac{S}{4T}\right)\right] + \log 1/u$$

If Q is constant, bracketed terms are constant and $(\log s)$ is related to $(\log t/r^2)$ in the same way as $(\log W(u))$ is related to $(\log 1/u)$.

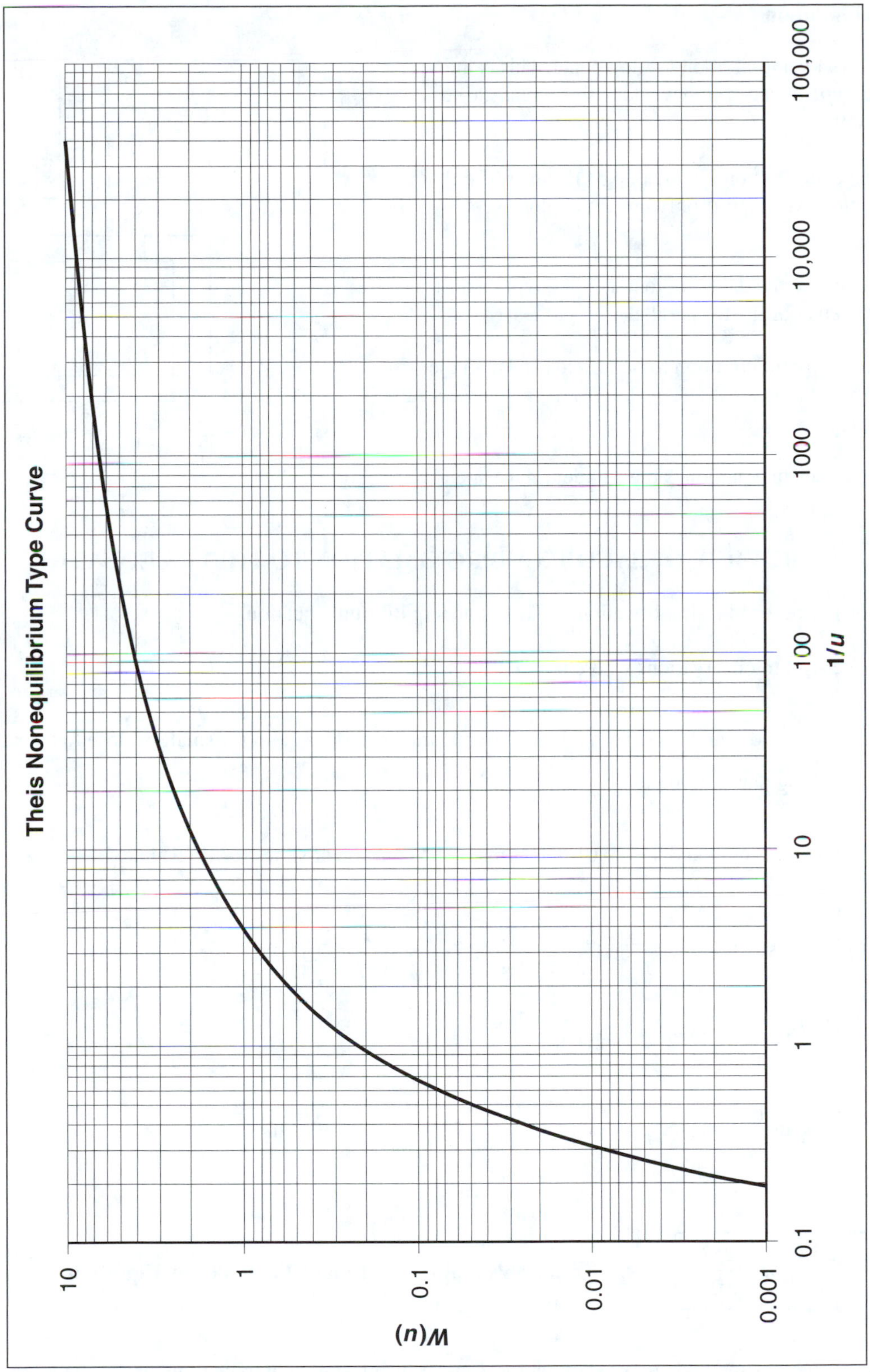

Figure 9.1—Theis-type curve for ideal confined aquifers.

Graphical Solution

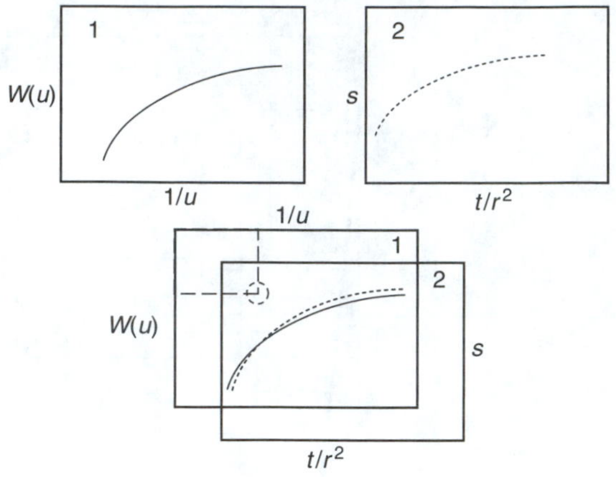

1. Plot $W(u)$ versus $1/u$ on log-log paper. This is the *Theis-type curve*. You have done this already, or use Figure 9.1.
2. Plot s versus t/r^2 on identical log-log paper (if only one well, plot s versus t; if only one time, plot s versus $1/r^2$).
3. Superimpose the data plot to match the type curve (you must keep coordinates parallel).
4. Select any point common to both fields and read the four values (e.g., at $W(u) = 1$ and $1/u = 1 = u$, read s and t/r^2).
5. Insert values in equations above to determine T and S.

REVIEW OF JACOB'S APPROXIMATION TO THEIS'S EQUATION

Jacob (1950) introduced a simplification of Theis's nonequilibrium equation.

Validity: Restrict to observations where $\boldsymbol{u \leq 0.01}$.

Because $u = \dfrac{r^2S}{4Tt}$ and S and T are formation constants, this method is valid for small r (close to the pumping well) and/or large t (long pumping times).

Basis: When r^2/t is small and $u \leq 0.01$, the terms in the convergent series for $W(u)$ may be ignored after the second term.

$$W(u) = [-.5772 - \ln u] \quad \text{and} \quad s = \frac{Q}{4\pi T}W(u) = \frac{Q}{4\pi T}(-.5772 - \ln u)$$

$$s = \frac{Q}{4\pi T}\left(\ln\frac{1}{u} - .5772\right), \quad \text{and since} \quad e^{.5772} = 1.78$$

$$s = \frac{Q}{4\pi T}\left(\ln\frac{1}{u} - \ln 1.78\right) = \frac{Q}{4\pi T}\left(\ln\frac{4Tt}{r^2S}/1.78\right) \qquad s = \frac{Q}{4\pi T}\ln\frac{2.25Tt}{r^2S},$$

or

$$\boxed{s = \frac{2.3Q}{4\pi T}\log\frac{2.25Tt}{r^2S}}$$

Jacob's Approximation of Theis's Flow Equation
(for $u \leq 0.01$)

Application: Jacob's method allows solution for T and S, using simple graphical methods, by plotting

1. s versus $\log t$ for one observation well, at various times
2. s versus $\log r$ for several observation wells, at one time
3. s versus $\log t/r^2$ for several observation wells, at various times

Jacob's Method (1) *s* as $f(t)$

$$\text{at } t_1, \quad s_1 = \frac{2.3Q}{4\pi T}\log\frac{2.25Tt_1}{r^2S}; \quad \text{at } t_2, \quad s_2 = \frac{2.3Q}{4\pi T}\log\frac{2.25Tt_2}{r^2S}$$

$$\Delta s = s_2 - s_1 = \frac{2.3Q}{4\pi T}(\log Ct_2 - \log Ct_1), \quad \text{where} \quad C = \frac{2.25T}{r^2S}$$

$$\Delta s = \frac{2.3Q}{4\pi T}\log(t_2/t_1)$$

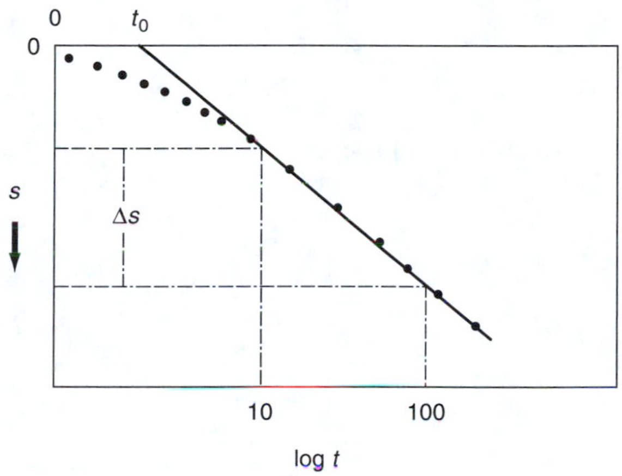

1. Plot s versus $\log t$.
2. Draw straight-line portion to
   (a) extend through one log cycle
   (b) project to $s = 0$
3. Select one log cycle of t.

   Note: $\log\left(\frac{10t}{t}\right) = 1$
4. Read Δs over log cycle.

$$\Delta s = \frac{2.3Q}{4\pi T}$$

To determine *T*: $$\boxed{T = \frac{2.3Q}{4\pi\Delta s}}$$

To determine *S*: for $s = 0$, $\frac{2.3Q}{4\pi T}\log\frac{2.25Tt_0}{r^2S} = 0$

$$\log\frac{2.25Tt_0}{r^2S} = 0 \qquad \frac{2.25Tt_0}{r^2S} = 1$$

$$\boxed{S = \frac{2.25Tt_0}{r^2}}$$

Jacob's Method (2) *s* as $f(r)$

Approach is the same.

$$s = s_2 - s_1 = \frac{2.3Q}{4\pi T}(\log C/r_2^2 - \log C/r_1^2), \quad \text{where} \quad C = \frac{2.25}{S}$$

$$\Delta s = \frac{2.3Q}{4\pi T}\log\frac{r_1^2}{r_2^2} = \frac{2.3Q}{4\pi T} \times 2\log\frac{r_1}{r_2} \qquad \Delta s = \frac{2.3Q}{2\pi T}\log r_1/r_2$$

1. Plot s versus $\log r$.

2. Draw straight-line portion through one cycle, to $s = 0$.

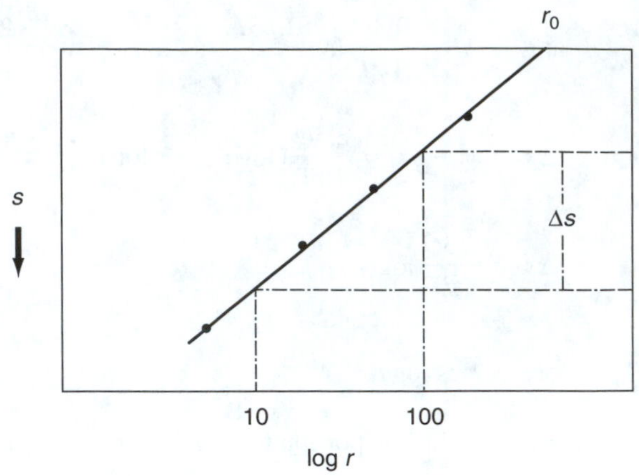

3. Select log cycle.

$$\left(\log \frac{r}{10r} = -1\right)$$

4. Read Δs over log cycle.

$$\Delta s = \frac{-2.3Q}{2\pi T}$$

To determine *T*: $$T = \frac{-2.3Q}{2\pi \Delta s}$$

To determine *S*: $$S = \frac{2.25Tt}{r_0^2}$$

Jacob's Method (3) *s* as $f(t/r^2)$

Similar to above, plot s versus $\log t/r^2$

$$T = \frac{2.3Q}{4\pi \Delta s} \qquad S = 2.25T\left(\frac{t}{r^2}\right)_0$$

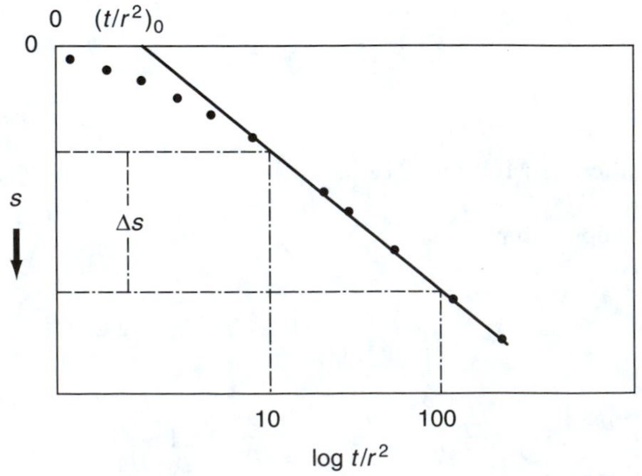

ANALYSIS OF AQUIFER TEST DATA

Analysis of Aquifer Test Data by Theis and Jacob Methods

1. A new well has been drilled in order to test an aquifer. It was pumped at a rate of 220 gpm, while an observation well 824 feet away was used to observe drawdown. The data obtained are summarized in the following table.

Drawdown (ft)	Time (min)	Time (days)
0.40	3.5	2.39×10^{-3}
0.74	5.0	3.47×10^{-3}
1.00	6.2	4.30×10^{-3}
1.35	8.0	5.55×10^{-3}
1.55	9.2	6.40×10^{-3}
2.08	12.4	8.60×10^{-3}
2.70	16.5	1.15×10^{-2}
3.02	20	1.39×10^{-2}
3.95	30	2.09×10^{-2}
5.71	60	4.17×10^{-2}
7.03	100	6.94×10^{-2}
8.45	200	1.39×10^{-1}
9.87	320	2.23×10^{-1}
10.20	380	2.64×10^{-1}
10.98	500	3.47×10^{-1}

(a) Compute the aquifer constants, using the Theis method.

T ____________________

S ____________________

(b) Is the aquifer confined or unconfined?

(c) Compute the aquifer constants, using the Jacob method.

T ____________________

S ____________________

Is the solution valid?

(d) Explain any differences. Which of the two methods gives more accurate results in this case? Why?

(e) Under what conditions of testing should the two methods agree well (for this aquifer test)?

Analysis of Aquifer Test Data by Jacob Method—Several Observation Wells at One Time

2. A well has been pumping 600 gpm for 4 hours. Drawdowns were observed at three observation wells.

Observation Well A:	distance = 160 ft	Drawdown = 4.70 ft
Observation Well B:	distance = 400 ft	Drawdown = 3.50 ft
Observation Well C:	distance = 1200 ft	Drawdown = 2.00 ft

(a) Determine the aquifer constants, using Jacob's method.

T ____________________

S ____________________

(b) Is the aquifer confined or unconfined?

(c) Is the solution valid?

Analysis of Aquifer Test Data—Your Choice of Method

3. A shallow test well at a proposed dam site was pumped at 1100 gpm. An observation well 16 feet away showed these observed drawdowns:

Drawdown	Time	
(ft)	(min)	(days)
0.96	4	2.8×10^{-3}
1.26	10	7.0×10^{-3}
1.58	25	1.7×10^{-2}
1.79	40	2.8×10^{-2}
2.01	75	5.2×10^{-2}
2.39	230	1.6×10^{-1}
2.67	500	3.5×10^{-1}
3.08	1600	1.1

(a) Does the dam site have a permeable surface (i.e., does the well pump from an unconfined aquifer)?

(b) Will seepage beneath the dam be significant (i.e., does the material have a high transmissivity)?

(c) Discuss your results, especially the accuracy of your results. You may first want to review the basic theory underlying Theis's solution (and, therefore, Jacob's approximation solution, as well). For what kind of aquifer is the method intended? Are Theis's assumptions valid in this case?

Analysis of Aquifer Test Data—Computer Curve-Matching Program

4. Rerun the data from Problem 1 using the THCVFIT [THeis CurVe FIT] program. This is an old DOS program designed to do what you have done manually—fit data points to the Theis-type curve, pick common coordinates, and calculate formation constants. The program, in both BASIC and compiled forms, is on your CD in the directory THCVFIT, along with the 15 data points from Problem 1 in the file LAB9PRO1.DAT.

   Note that this software is provided without any restriction on its use, duplication, and redistribution, but it is no longer supported.

   (a) In a Windows environment:

   START

   PROGRAMS

   MSDOS Command Prompt

   a: (or other designation for your CD drive)

   cd thcvfit

   thcvfit

   when opening screen appears

   <Enter>

   You can import the data points directly by using the file LAB9PRO1.DAT, or you can key in data points manually.

   When moving your data plot to match the type curve, use the cursor keys on the numeric keypad (turn off the Number Lock, if on, by hitting the NumLock key).

   T ____________________

   S ____________________

   (b) Compare these results with the formation constants you determined manually. Which is more accurate? Why?

   (c) You can also use this program now to check your other problem solutions. If you have access to AQTESOLV (e.g., included with Fetter's 3rd and 4th editions), use this to analyze the data in Problem 3. Compare and discuss the results. Which is more accurate? Why?

Predicting Response (Drawdown) of a Known Aquifer to Pumping

5. A confined aquifer has a transmissivity of 10,000 gpd/ft and a storativity of 2.4×10^{-5}. If a well were installed in this aquifer and pumped continuously at 100 gpm, what would be the effect on a well 1 mile away after one year?

Solution: Up to now, you have used Theis's method (and Jacob's approximation) to determine aquifer constants by observing drawdown as a function of time and distance. You can use the same techniques in reverse to predict drawdown as a function of time and distance if you know the aquifer properties, S *and* T.

Calculate $1/u$, *find* $W(u)$, *and calculate* s.

LAB 10

AQUIFER TESTING II: NONIDEAL AQUIFERS

PURPOSE: Become more familiar with aquifer test data and methods of interpretation, especially for data that do not conform to the type curve for ideal, confined aquifers.

OBJECTIVES: Learn to interpret drawdown data that do not fit the ideal Theis-type curve used in the last lab.

Learn to interpret drawdown data that do not fit the ideal Jacob-solution model.

Learn to recognize certain curve shapes for nonideal aquifers.

Recognize that one can sometimes estimate aquifer conditions by visually examining test data.

READING ASSIGNMENT: Read material in your textbook that covers aquifer tests, especially for semiconfined aquifers, unconfined aquifers, and aquifers of limited extent (e.g., Fetter, 2001, pp. 177–188, 208–209).

INTRODUCTION

In this lab, one of the things you will be asked to do is *estimate* the aquifer conditions—whether confined or unconfined—by inspecting the drawdown data. This is difficult to do, especially the first few times you try, but go for it—you have nothing to lose. You might even find that you can get a bit of a feel for it and make some intelligent guesses.

When you are presented (as you have been) with tabular data of drawdown versus time:

1. Consider the distance from the observation well to the pumping well, the pumping rate, the test duration, and the total drawdown. Do you think the aquifer would respond like this if it were confined? Unconfined?

2. How soon does the observation well show a response? Considering the parameters above (1), is this an "instantaneous" response?

3. Look at the time intervals between measurements—are they fairly regular (though increasing)? Then look at the drawdown increments—are they fairly regular? Are there interruptions in the drawdown? Does drawdown, in fact, cease?

Try this approach with the data from Lab 9, Problem 3.

1. After pumping at more than 1000 gpm, for more than 24 hours, just 16 feet away the drawdown is only 3 feet. What do you think?

In this lab, you will use two families of curves that are departures from the ideal Theis-type curve. These curves are portions of plates published by S.W. Lohman (1979) in U.S. Geological Survey Professional Paper 708. It is worthwhile to purchase a copy (available at a nominal cost; the U.S. Government Printing Office reprints the paper regularly) so that the entire extent of the curves is available. The scale is the same as in this manual; if log-log paper isn't available at this scale, you can photocopy these type curves to match the scale of your paper, or you can use an overlay on the type curve.

Problem 1. The drawdown data below are from an observation well 1000 ft from a well pumping at a constant discharge of 1000 gpm.

Drawdown (ft)	Time (days)	Drawdown (ft)	Time (days)
0.00	1.30×10^{-4}	1.95	3.47×10^{-2}
.00	3.47×10^{-4}	2.08	6.94×10^{-2}
.02	6.94×10^{-4}	2.09	1.39×10^{-1}
.14	1.39×10^{-3}	2.09	3.47×10^{-1}
.55	3.47×10^{-3}	2.10	6.94×10^{-1}
.99	6.94×10^{-3}	2.11	2.30
1.46	1.39×10^{-2}		

(a) Consider the distance between the pumped well and the observation well, the pumping rate and duration, and the amount and rate of drawdown. *Estimate* what the aquifer conditions might be.

(b) Plot the data, log *s* versus log *t*.

(c) Overlay your data plot onto the Theis-type curve (Fig. 9.1) . Describe the fit.

(d) What do you think happened during the test?

(e) If you look at the type curves in Figures 10.2 (page 83) and 10.3 (page 85), you should be able to recognize this curve shape, at which point you can determine the aquifer constants.

T ______________________________

S ______________________________

(f) From the secondary curves, you can also derive additional information about the overlying bed—in this case, the leakance. If the driller's log shows the overlying bed to be 100 ft thick, what is its vertical hydraulic conductivity?

Leakance ______________________________

K' ______________________________

(g) Describe the aquifer.

Problem 2. An observation well is 73 ft from a well pumping at a constant rate of 1080 gpm. The following drawdowns were measured.

Drawdown (ft)	Time (min)	Drawdown (ft)	Time (min)
0.12	0.16	1.06	20
.26	.34	1.17	40
.43	.58	1.31	100
.64	1.00	1.65	300
.78	1.50	2.09	800
.94	3.0	2.27	1200
.99	6.0	2.59	2500
1.02	10		

(a) Considering the separation of the wells, the pumping rate, and the variation of drawdown with time, *estimate* what the aquifer conditions might be.

(b) Plot the data, log s versus log t.

(c) Overlay your data plot onto the Theis-type curve (Fig. 9.1) . Describe the fit.

(d) What do you think happened during the test?

(e) If you look at the type curves in Figures 10.2 and 10.3, you should be able to recognize this curve shape, at which point you can determine the aquifer constants.

T ______________________________

S ______________________________

(f) From the secondary curves, you can also derive additional information about the anisotropy of the aquifer. If the aquifer is 40 ft thick, describe the anisotropy.

K_h ______________________________

K_v ______________________________

(g) Describe the aquifer.

Problem 3. The drawdown data here are from an observation well 100 ft from a pumping well.

Drawdown (ft)	Time (min)
.03	30
.10	40
.14	50
.20	60
.22	70
.26	80
.30	90
.35	100
.42	120
.50	140
.65	200
.78	260
.90	350
1.02	410
1.20	500
1.50	700
1.82	1000
2.15	1500
2.42	2000
2.70	2500
3.30	4000
4.20	8000

(a) Consider the drawdown rate carefully. Can you suggest anything about the aquifer geometry? (This one is a bit subtle.)

(b) Plot the data, log s versus log t.

(c) Overlay your data plot onto the Theis-type curve (Fig. 9.1). Describe the fit.

(d) Is the drawdown considerably more or considerably less than expected? What do you think happened during the test?

(e) Plot the data, s versus log t (semilog plot). Does a pattern emerge?

(f) Refer to the section *Aquifer of Limited Extent* at the end of this lab.

(g) Describe fully the aquifer and the aquifer geometry.

Problem 4. Following are data from a pump test run after the completion of a 2-ft-diameter, 120-ft-deep residential well. The test was run pumping at 10 gpm.

Drawdown (ft)	Time (min)
0.21	0.5
.43	1.0
.85	2
1.69	4
3.38	8
6.34	15
12.7	30
25.4	60
50.7	120
76.6	180

(a) Describe the aquifer, and comment on the suitability of the well as a domestic supply for a family of four. (Appendix 2 offers guidelines on quantities required for water supply.)

(b) If you are confident in your interpretation of this well, go on to (c); if not, do the following exercise.

Pump from a full 55-gallon drum at 1 gpm, and calculate drawdown data.

Drawdown (ft)	Time (min)
	0.5
	1
	2
	4
	8
	15
	30
	45

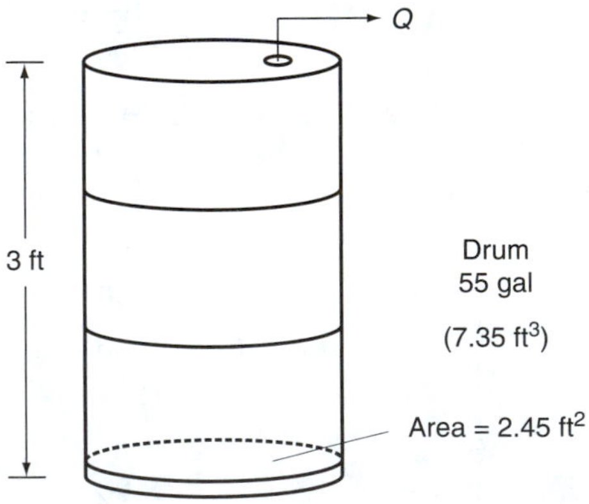

Compare these data with the well data above. What is your conclusion?

(c) Figure 10.1 is a copy of a well-completion report. These reports are a common source of groundwater information, usually kept on file by a state government agency (the state engineer's office in Colorado). The report is for a fairly typical well in the Colorado mountains, and it reports a sustained yield of 3 gpm. Is this well suitable for a domestic supply for a family of four?

Note: If one stops at this point, surely, 3 gpm is sufficient. The careful engineer or hydrogeologist, however, will consider well storage and do further calculations.

THIS FORM MUST BE SUBMITTED WITHIN 60 DAYS OF COMPLETION OF THE WORK DESCRIBED HERE-ON. TYPE OR PRINT IN BLACK INK.

COLORADO DIVISION OF WATER RESOURCES
1313 Sherman Street - Room 818
Denver, Colorado 80203

WELL COMPLETION AND PUMP INSTALLATION REPORT

PERMIT NUMBER 780413 C

WELL OWNER Y. B. DREIGH

ADDRESS COLORADO SPGS CO

DATE COMPLETED APRIL 9, 19 78

WELL LOG

From	To	Type and Color of Material	Water Loc.
0	5	OVERBURDEN — CLAY, SAND	
5	75	PINK GRANITE	
75	105	DARK GRANITE	
105	265	LIGHT GRAY GRANITE	
265	320	PINK GRANITE w/ WHITE LENSES	280

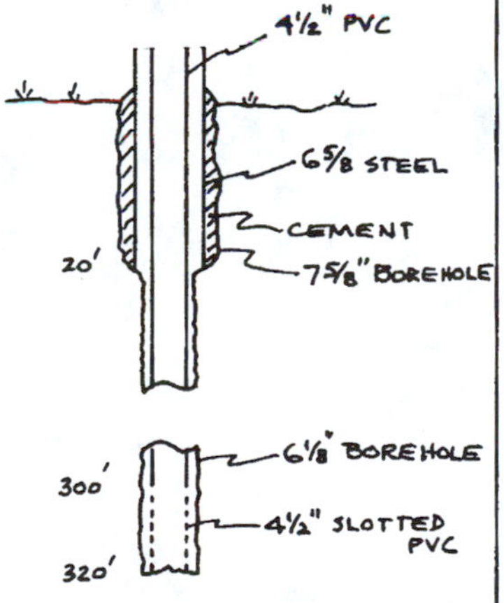

TOTAL DEPTH 320 FT

Use additional pages necessary to complete log.

NE ¼ of the NE ¼ of Sec. 14

T. 9 S, R. 74 W, 6TH P.M.

HOLE DIAMETER

7⅝ in. from 0 to 20 ft.

6⅛ in. from 20 to 320 ft.

___ in. from ___ to ___ ft.

DRILLING METHOD AIR PERCUSSION

CASING RECORD: Plain Casing

Size 6⅝ & kind STEEL from +1 to 20 ft.

Size 4½ & kind PVC from 10 to 300 ft.

Size ___ & kind ___ from ___ to ___ ft.

Perforated Casing

Size 4½ & kind SLOT from 300 to 320 ft.

Size ___ & kind ___ from ___ to ___ ft.

Size ___ & kind ___ from ___ to ___ ft.

GROUTING RECORD

Material CEMENT

Intervals 20 FT

Placement Method POURED

TEST DATA

Date Tested APRIL 10, 19 78

Static Water Level Prior to Test 110 ft.

Type of Test Pump AIR

Length of Test 2 HOURS

Sustained Yield (Metered) 3 GPM

Final Pumping Water Level 320

Figure 10.1—Well-completion report. This is a copy of an actual report, although the name and location are fictitious.

SEMICONFINED AQUIFERS

The type curves in Figure 10.2 on page 83 are used with nonsteady flow in an aquifer that is confined by a semipermeable layer. The use of these curves is similar to the Theis procedure you have learned already—observational data are plotted on same-scale log-log tracing paper and fit to the type curves.

Additional information is also available about the semiconfining layer—namely, its *leakance*, a measure of how much the semiconfining layer will leak. Leakance is defined as K'/b', where K' is the vertical hydraulic conductivity of the semiconfining layer and b' is its thickness.

The governing equations for these curves are:

$$s = \frac{Q}{4\pi T} L(u, v) \quad \text{where} \quad v = \frac{r}{2}\left(\frac{K'}{b'T}\right)^{1/2}$$

and $L(u, v)$ is the *leakance function of* u *and* v;

from which

$$T = \frac{Q}{4\pi s} L(u, v)$$

$$S = 4T \frac{t/r^2}{1/u}$$

$$\frac{K'}{b'} = 4T \frac{v^2}{r^2}$$

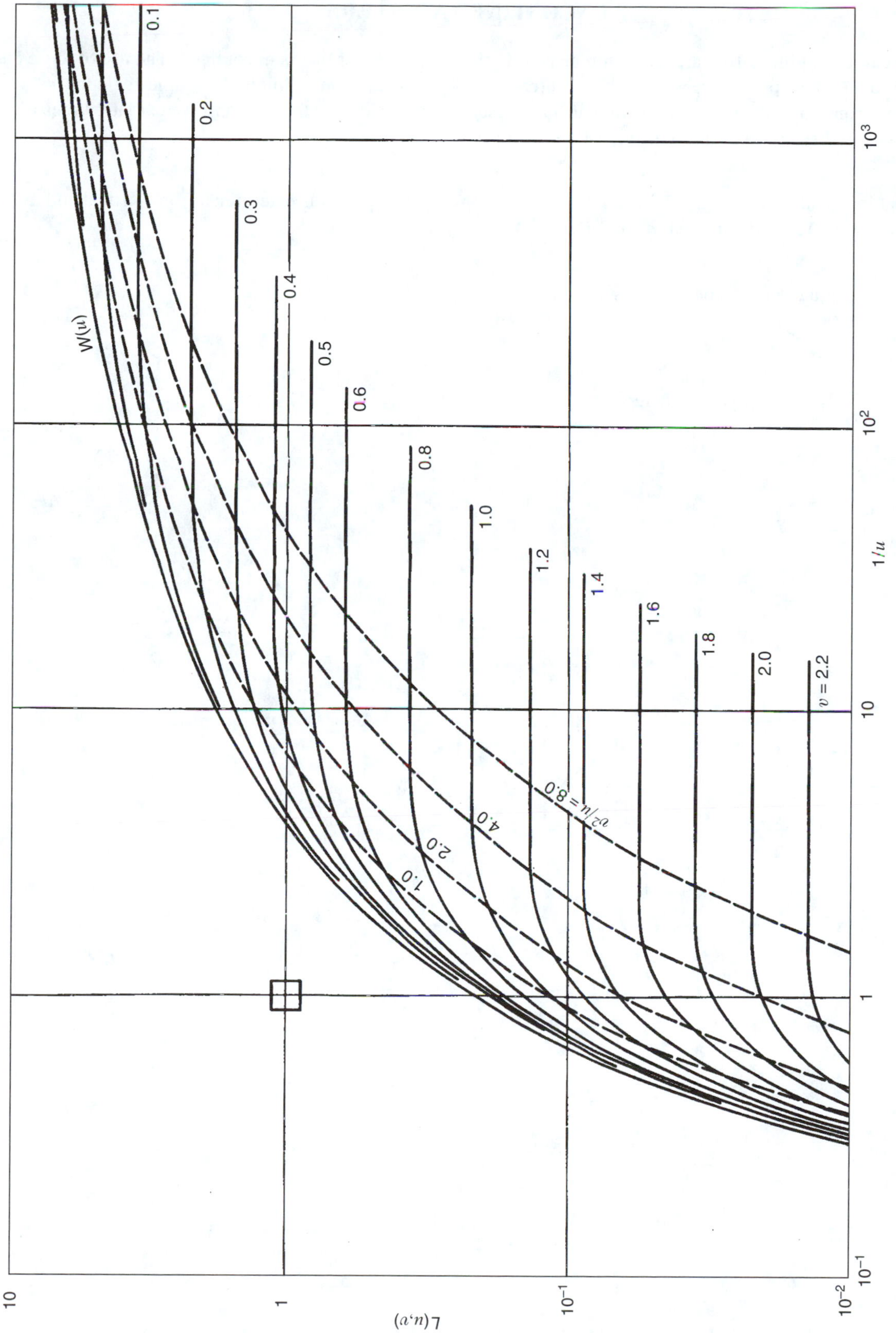

Figure 10.2—Type curves for nonsteady, radial, semiconfined flow (from Lohman, 1979, Plate 3A).

UNCONFINED AQUIFERS

The type curves in Figure 10.3 are used with nonsteady flow in an aquifer that is unconfined. The use of these curves is similar to the Theis procedure you have used already—observational data are plotted on same-scale log-log tracing paper and fit to the type curves. In matching these data, however, use the data observed at later times (data plotted to the right) and fit them to the type-B curves (also to the right).

Additional information is also available about the anisotropy of the aquifer, as determined from vertical and horizontal hydraulic conductivities, K_v and K_h.

The governing equations for these curves are:

$$s = \frac{Q}{4\pi T} W(u_A, u_B, \beta) \quad \text{where} \quad \beta = \frac{r^2 K_v}{b^2 K_h}$$

from which,

$$T = \frac{Q}{4\pi s} W(u_A, u_B, \beta)$$

$$S_y = \frac{4Tt}{r^2} u_B$$

$$K_h = \frac{T}{b}$$

$$K_v = \frac{K_h b^2 \beta}{r^2}$$

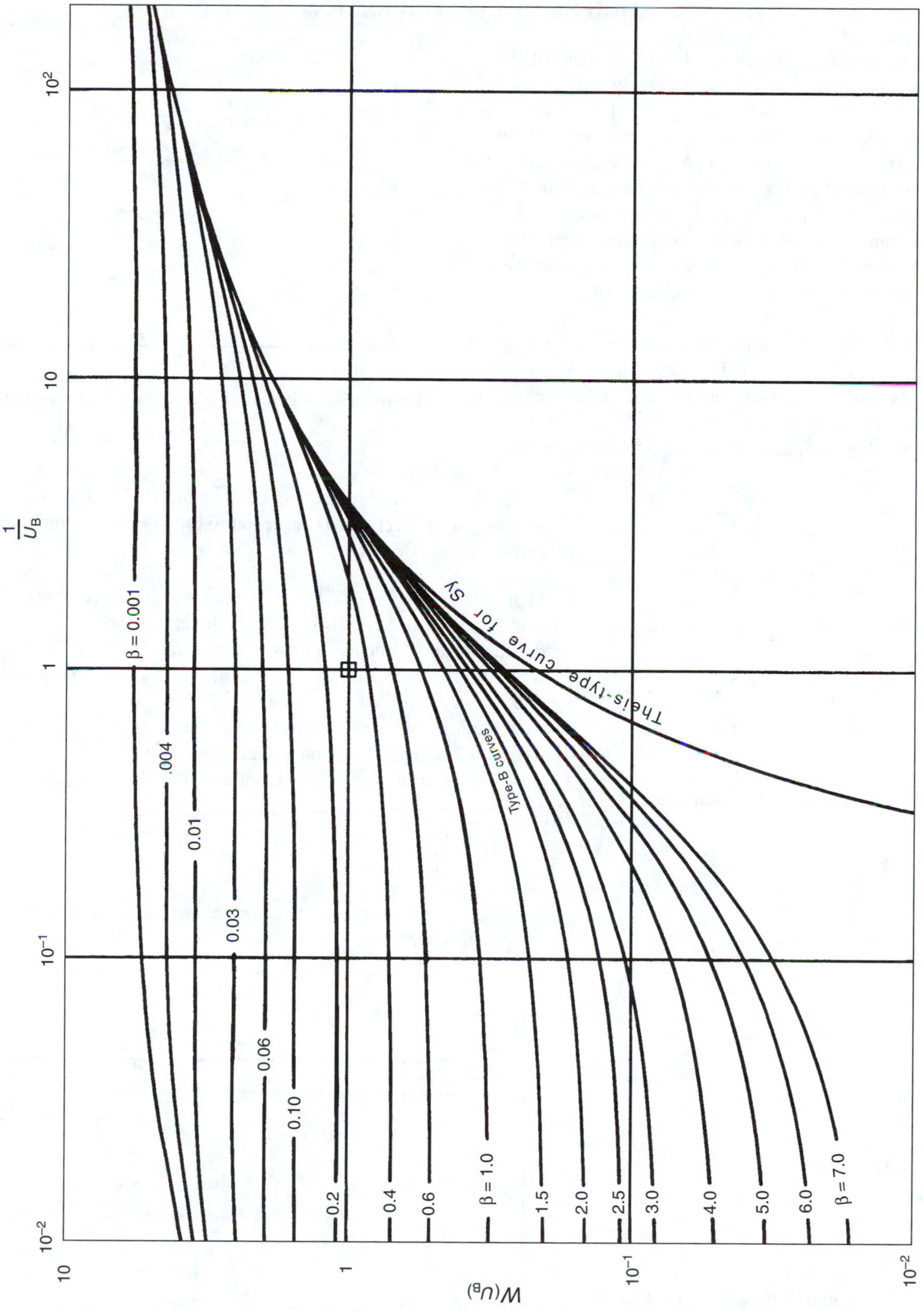

Figure 10.3—Type curves for nonsteady, radial, unconfined flow (modified from Neuman, 1975, Figure 1).

AQUIFER OF LIMITED EXTENT

A well pumping from an aquifer of infinite extent will develop a cone of depression that will grow indefinitely. As you saw in the previous lab, on a log-drawdown versus log-time graph, this will be the Theis type curve. Any departure from the type curve indicates that a hydrogeologic boundary has been reached by the growing drawdown cone (or, as seen earlier, that induced recharge is occurring through a semiconfining layer). Boundaries can be considered of two types: recharge boundaries, like surface water-bodies, and barrier boundaries, like impermeable materials.

Figure 10.4 illustrates the effect of boundaries on a log-log plot. Sometimes the effects of boundaries are a bit subtle to see on these plots, especially at long pumping times when the slope is fairly flat. These effects usually can be better observed on Jacob-type plots because they form departures from a straight line.

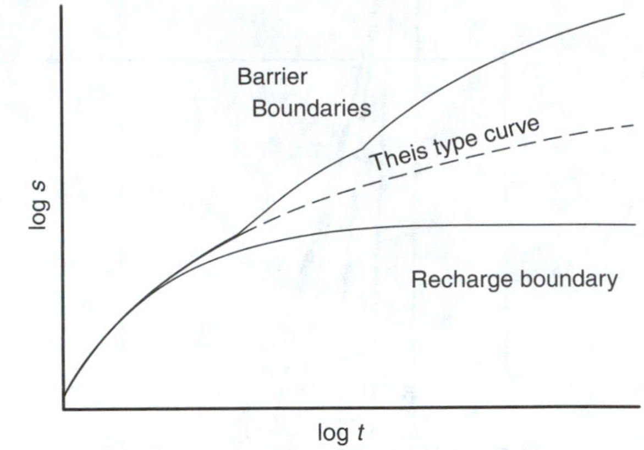

Figure 10.4—Effect of boundaries on drawdown.

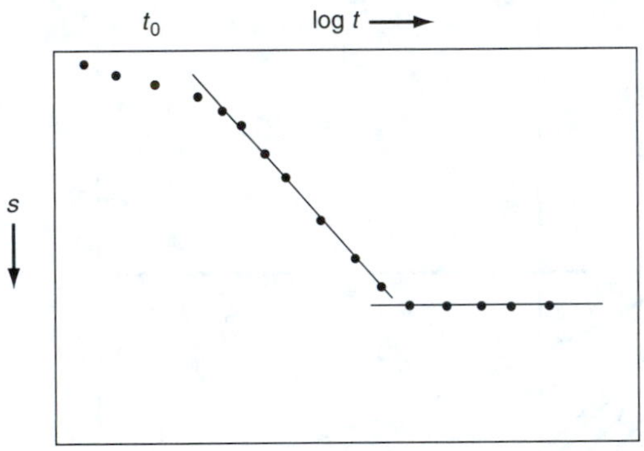

Figure 10.5—Recharge boundary.

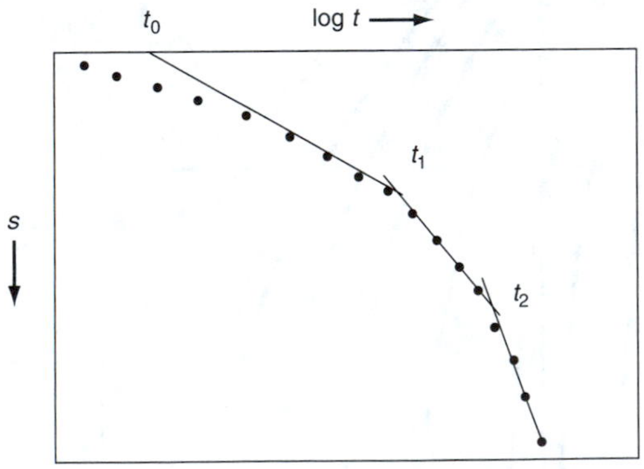

Figure 10.6—Barrier boundaries.

The cross section in Figure 10.7 illustrates the geometry of the wells corresponding to the drawdown shown in Figure 10.6.

Figure 10.5 illustrates the effect of a recharge boundary on a pumping well. Recharge is sufficient to provide flow to the well, and drawdown ceases.

Figure 10.6 shows the effect of two barrier boundaries. At time t_1, the cone has reached the first barrier. At this time, the slope of the drawdown curve doubles—that is, the slope after t_1 is twice the slope before t_1. This is because the effect of the barrier is the same as the effect of an image well pumping at exactly the same rate as the real well. Similarly, when the second barrier is reached, the slope becomes three times as great, representing the real well and two image wells. The distance to the wells can be calculated from the relationship

$$\frac{t_0}{r_0^2} = \frac{t_1}{r_1^2} = \frac{t_2}{r_2^2}$$

where radial distances refer to distances measured from the *observation* well.

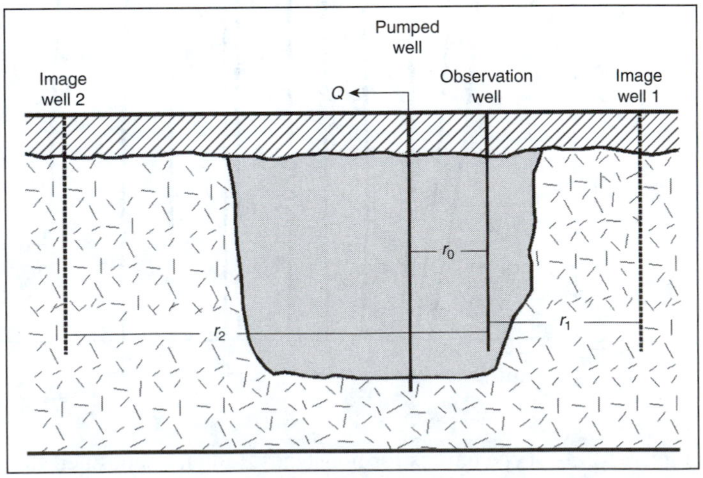

Figure 10.7—Aquifer with two impermeable boundaries.

LAB 11

AQUIFER TESTING III: SLUG-TEST DATA EVALUATION

PURPOSE: Introduce common slug-test analysis techniques and their assumptions/limitations.

OBJECTIVES: Evaluate slug-test data using the Hvorslev (1951) graphical method.

Evaluate slug-test data using the Bouwer and Rice (1976) method.

Understand limitations of slug-test analysis techniques.

READING ASSIGNMENT: Read material in your textbook that covers slug tests (e.g., Fetter, 2001, Section 5.6) and hydraulic conductivity ranges of unconsolidated aquifer materials (e.g., Fetter, 2001, Table 3.7).

MATERIALS: Three-cycle semilog paper (Problem 1) and 4-cycle semilog paper (Problem 2), ruler, and calculator. Spreadsheets can also be used to make semilog plots.

INTRODUCTION

The slug test, an alternative to full-scale aquifer testing, has become common practice in hydrogeologic investigations. Slug testing involves inducing a hydraulic gradient in a well by instantaneously adding or removing a slug (i.e., a small volume of water or a solid cylinder). The water level in the well will eventually return to the water level in the surrounding aquifer before the slug was introduced. The rate of the return of the water level to the steady value is proportional to the hydraulic conductivity. The water level is recorded over time, and hydraulic conductivity of the aquifer materials near the well can be estimated by analyzing the change in head as a function of time, using various analytical methods.

Slug tests are inexpensive and convenient to perform because they do not require elaborate pumping procedures, and they can be performed in contaminated aquifers without withdrawing contaminated groundwater. Simplistic and widely applicable slug-test analysis techniques have been shown to produce accurate estimates of hydraulic conductivity. The two methods treated here are widely used for both unconfined and confined materials, although there are some limitations on their applicability (see, for example, Dawson and Istok, 1991).

Problem 1. Students at Colorado School of Mines (CSM) conducted a slug test in a 2-in.-diameter well that was installed in relatively homogeneous, isotropic, silty sand using an 8-in. outer diameter hollow-stem auger. The well was installed to a total depth of 40 feet below ground surface (bgs), and the well is screened from 29.5 ft to 39.5 ft bgs. The screened section of the well is surrounded by a gravel pack of 10–20 mesh-size sand from 28.5 ft to 40 ft bgs. Field observations made by the site geologist during well installation indicate that the well is installed in an unconfined aquifer.

The following measurements were recorded in the field prior to conducting a slug test.

Height of well casing above ground surface:	2 ft
Initial depth to water (below top of casing [btoc]):	8.15 ft

(a) The Hvorslev equation can be applied to a piezometer test if the length of the well screen (L_e) is more than eight times the radius of the well screen (R) and the aquifer is unconfined.

$$K = \frac{r^2 \ln(L_e/R)}{2L_e t_{37}}$$

What is the ratio of L_e/R for this problem? Can the Hvorslev method be applied to this problem?
A falling-head slug test was performed by adding a small volume of water to the well and manually recording the depth to water as a function of time. The depths-to-water data are shown below (measured below top of casing [btoc]).

Time (sec)	**dtw[1] (ft)**	***h* (ft)**	**h/h_0**	**Time (sec)**	**dtw[1] (ft)**	***h* (ft)**	**h/h_0**
0	6.85			165	7.88		
10	6.95			180	7.91		
20	7.03			195	7.95		
30	7.12			225	8.00		
45	7.25			255	8.05		
60	7.35			285	8.07		
75	7.45			315	8.09		
90	7.55			345	8.10		
105	7.65			375	8.12		
120	7.73			405	8.13		
135	7.80			435	8.14		
150	7.84			465	8.15		

[1]dtw = depth to water in well

(b) A water-level change of 4 inches to 20 inches is believed to produce the best results for slug tests performed in small-diameter wells. How many inches did the water level rise during this test? Based on this information, should the analysis yield a reliable estimate of hydraulic conductivity?

(c) Compute the maximum height (h_0) to which the water level rises above the static water level, and compute the height (h) of the water level above the static water level for each measurement time.

(d) Compute the head ratio (h/h_0) for each measurement time.

(e) Plot the head ratio (h/h_0) on a log scale versus time (t) on an arithmetic scale.

(f) Draw a line of best fit through the portion of the graph that best characterizes the hydraulic conductivity of the aquifer.

(g) At what time (t_{37}) does the water level fall to 37 percent of the initial change?

(h) Calculate the hydraulic conductivity (K) of the aquifer using the Hvorslev equation (cm/sec).

(i) Compare the results of your calculation of K to the range of hydraulic conductivity values given in your hydrogeology textbook. Does your calculated value for K seem reasonable?

Problem 2. In this problem, another well is used to conduct a slug test. You are given the well-construction details, and you know that the lower boundary of the aquifer is 59 feet below the ground surface. A 2-in. inner-diameter well was installed using an 8-in. outer-diameter hollow-stem auger. The partially penetrating well was installed to a total depth of 45 feet below ground surface (bgs), and the well is screened from 34.5 feet bgs to 44.5 feet bgs. The screened section of the well is surrounded by a gravel pack of 10–20 mesh-size sand. The well bore above and below the screen is sealed off, such that the gravel pack essentially surrounds only the screen.

The following measurements were recorded in the field prior to conducting a slug test.

Height of well casing above ground surface:	2 ft
Initial depth to water (below top of casing [btoc]):	29.20 ft

A rising-head slug test was performed by removing approximately one-half gallon of water from the well and recording the height of the water column above a reference, using a pressure transducer, as a function of time. The pressure measurements were electronically recorded using a data logger. These measurements have been corrected to give the displacement (absolute value) of the water level from static conditions, H_t (i.e., height of the water in the well below the initial hydrostatic water level before the slug was removed). Transducer readings are normally collected about every second. For the practical purposes of this laboratory, a subset of the entire data set is provided below.

Time (sec)	H_t (ft)	Time (sec)	H_t (ft)
0	3.065	70	0.063
5	2.239	80	0.039
10	1.778	90	0.026
15	1.230	100	0.019
20	0.910	110	0.014
25	0.708	120	0.012
30	0.513	130	0.011
40	0.302	140	0.011
50	0.166	150	0.010
60	0.104	160	0.010

(a) Determine the saturated thickness of the aquifer (h) at the location of this well.

(b) Calculate the height of the static water level (L_w) above the bottom of the well screen.

(c) Compute $\ln(R_e/R)$ by utilizing well construction information and aquifer details, using the following equation:

$$\ln\frac{R_e}{R} = \left[\frac{1.1}{\ln(L_W/R)} + \frac{A + B\ln[(h - L_W)/R]}{L_e/R}\right]^{-1}$$

where A and B are dimensionless parameters obtained from the plot of A and B versus L_e/R in your textbook, or in the paper by Bouwer (1989).

(d) Under what well-geometry and hydraulic conditions can this equation be used?

(e) Plot the displacement (H_t) on a log scale versus time (t) on an arithmetic scale.

(f) Draw a line of best fit through the portion of the graph that best characterizes the hydraulic conductivity of the aquifer.

(g) From the graph, determine displacement (H_0) at time $t = 0$ and a displacement (H_t) at a time t. Choose a time when H_t lies on the straight-line portion of the graph.

(h) Calculate the hydraulic conductivity (K) of the aquifer, using the Bouwer and Rice (1976) equation (use units of cm/sec).

$$K = \frac{r_c^2 \ln(R_e/R)}{2L_e} \ln\left(\frac{H_0}{H_t}\right)\frac{1}{t}$$

(i) Compare the results of your calculation of K to the range of hydraulic conductivity values given in your textbook. In what type of material do you think the well is completed?

(j) Suppose the plot from procedure (f) indicated a straight line at early times, another straight line at later times, and a nonlinear curve at the final times. Which data would you use for the analysis? Why?

(k) According to Bouwer and Rice (1976), the hydraulic conductivity estimates obtained by using their method should contain a percentage error ranging from 10–25 percent. It has been suggested that the Bouwer and Rice method tends to underestimate K values.

Provide a range for the true value of K based on this information.

LAB 12

AQUIFER TESTING IV: FIELD OBSERVATION OF AN AQUIFER TEST

PURPOSE: The purpose of this field lab is to acquaint you with procedures used in conducting an aquifer test (also known as a *pump test*).

OBJECTIVES: Witness the field activities at the beginning of an aquifer test.

Observe instruments and equipment used.

Observe procedures for measurement of drawdown as a function of time.

Observe procedures for measurement of pump discharge as a function of time.

Gain an appreciation for the accuracy and precision of typical aquifer test data.

NOTE: We will try to give you the opportunity to observe (at least the beginning of) an aquifer test, somewhere in the local area.

We will advise you of the time and location of the test with as much lead time as possible.

Generally, we cannot control exactly when a pump test will start. This lab, therefore, is optional, in that you may or may not be able to attend. Most tests run long enough that, even if you miss the beginning of the test, you should be able to catch some part of it.

READING ASSIGNMENT: Read the section in your textbook on conducting aquifer tests (e.g., Fetter, 2001, pp. 210–213).

PROCEDURES: Drive to the site of the test, where an instructor will meet you. Take a clipboard and your calculator. Wear normal field clothes.

Complete the form on the next page, and answer as many of the questions as you can.

AQUIFER TEST—FIELD OBSERVATIONS

Date: ________________ Organization/individual conducting test: ____________________________________

Location: ________________________________ Well ID: ___________________________________

________________________________ Well type: ___________________________________

Brief history of well:

Purpose of test:

Aquifer: ________________________________ Depth interval tested: _________________________

Describe or sketch map of pumping well and observation well(s):

Describe the type of equipment used:

How is drawdown measured? What do you think the accuracy is? Precision?

How is the pumping rate measured? Accuracy? Precision?

If you are able to do so without interfering with the test, get sufficient information to *predict* the testing results.

If you are able to do so without interfering with the test, record enough of the data to do an analysis of the data.

LAB 13

CONTAMINANT TRANSPORT I: GROUNDWATER AND CONTAMINANT VELOCITIES FROM BREAKTHOUGH CURVE ANALYSIS

PURPOSE: Become familiar with estimating contaminant transport model parameters.

OBJECTIVES: Determine input parameters for a mathematical contaminant transport model, using data from a laboratory tracer test.

Use the temporal method-of-moments to analyze breakthrough curve data to estimate average linear groundwater velocity, contaminant velocity, and retardation.

Estimate the longitudinal dispersion coefficient.

Estimate a biological degradation (decay) constant.

Use these data to predict the time for a contaminant to travel through an aquifer and the associated peak concentrations.

OVERVIEW: Transport of contaminants in groundwater is modeled to determine transport velocity and contaminant-plume shape. The models used for this, including simpler analytical models and more complex numerical models, require input parameters that characterize both the aquifer (hydraulic gradient, hydraulic conductivity, porosity, dispersivity, and heterogeneity) and the contaminant (degradation and retardation). Modeling success, of course, depends on the accuracy of these input parameters. Three approaches may be used: the input parameters may be estimated from a general knowledge of the aquifer (the least desirable approach), the parameters may be estimated from laboratory analyses of aquifer samples, or the parameters may be estimated from field tests in the aquifer itself, (e.g. using tracers).

Labs 13, 14, and 15 deal with contaminant transport.

Lab 13 deals with the estimation of input parameters from lab analyses of aquifer samples. You will not perform these tests; rather, you will analyze the results from such tests.

Lab 14 estimates the parameters by analysis of a tracer injected into the aquifer and uses the derived parameters in a simple analytical model to estimate travel of a contaminant plume.

Lab 15 models a contaminant plume and investigates the difference between assuming aquifer homogeneity, using a lumped-dispersion-coefficient approach, and modeling aquifer heterogeneity.

REQUIRED MATERIALS

PC computer

Spreadsheet software (e.g., Microsoft Excel)

INTRODUCTION

Problem Statement

Organic chemical waste leaked into an aquifer. You need to estimate the time required for the contaminant to travel from the source of contamination to a small stream. You also need to estimate the peak concentrations in the groundwater before it flows into the stream.

Scenario

A drum of dissolved chemical waste at a research facility in Florida leaked approximately 450 L. The contaminant was trichloroethene, or TCE. From the logs for the waste drum, the estimated concentration of TCE in the drum was 95 mg/L. Based on reports from the night crew, the leak probably occurred over a period of 1 day. By looking at the wetted ground surface, you estimate that the spill occurred over an area of about 3 m^2. Fortunately, groundwater tests had been conducted at a nearby site, so significant information about the aquifer at the spill site is available.

The water table is shallow and was estimated to be 20–40 cm below the ground surface at the spill. The groundwater zone is about 3.5 m thick and is bounded on the bottom by thick clay. The soil in the groundwater zone is loamy sand with a porosity of 0.36. The average linear groundwater velocity toward the stream is estimated to be 11 cm/day. The stream is about 100 m away, and there is concern that the detached TCE plume could enter the stream, a potential legal problem. As an initial assessment you may assume that the spill is nearly instantaneous, and the TCE was uniformly distributed down through the aquifer (z-direction) in one day. This problem may thus be approximated as a two-dimensional (x, y) contaminant transport problem with a pulse (or slug) input of dissolved contaminant that has an initial concentration (C_0) of 95 mg/L.

Our Approach to Solving the Problem

For this type of assessment, groundwater engineers often construct numerical or analytical models of contaminant transport. Simple analytical models are often used as a first estimate of transport characteristics before incurring the time and expense of more complex models. Of course, modelers must determine relevant transport parameters for model input. Examples of such model parameters are groundwater velocity, contaminant retardation coefficient, biochemical decay rate of the contaminant, and dispersivity values or dispersion coefficients. We may determine these parameters in the field and/or laboratory (with some obvious constraints), using hydraulic tracer tests. During a tracer test, a chemical of known concentration is injected into the soil and samples are measured at downstream locations. For this lab, a laboratory tracer test is used as a surrogate for a more expensive field test (a common practice). A field-testing method is treated in Laboratory 14. Normally, it would be best to conduct several of these laboratory-scale tests using soil from various locations at the site.

Here, we are attempting to mimic field conditions in a laboratory soil column to obtain reasonable values for the retardation coefficient and dispersivity. The data you will use to obtain estimates for these parameters are concentrations measured at the effluent of a laboratory soil column over time (Fig. 13.1 on page 96). The concentration is plotted versus time, and a breakthrough curve (BTC) plot is constructed.

To determine contaminant transport characteristics, we generally inject two tracers: the contaminant itself, in this case TCE; and a conservative tracer, in this case bromide. Conservative tracers are chemicals that do not react with the soil or groundwater system and experience negligible mass loss (due to decay or biodegradation, for example) during the duration of the tracer test. A conservative tracer travels with the groundwater at the same velocity as the groundwater.

Conservative tracer tests may also be conducted at a field site to determine the direction and velocity of the natural groundwater. However, these are considerably more difficult and expensive than a soil-column test. In addition, state and local regulatory agencies will seldom allow geological engineers and hydrogeologists to inject dissolved contaminants into the groundwater. Thus, as a first estimate of contaminant retardation and decay rates, tracer tests are often conducted in laboratory columns using samples of aquifer materials from the site.

BACKGROUND INFORMATION

The governing equation for transport of a sorbing chemical that experiences biochemical decay through a one-dimensional soil column is the advection-dispersion equation (ADE):

$$R\frac{\partial C}{\partial t} = -v\frac{\partial C}{\partial x} + D\frac{\partial^2 C}{\partial x^2} - \lambda C \tag{1a}$$

C is aqueous-phase contaminant concentration (M/L^3), v is the tracer velocity (L/T), D is the dispersion coefficient (L^2/T), R is the retardation coefficient (dimensionless) for linear, reversible, equilibrium sorption, and λ is the first-order biochemical decay constant (T^{-1}). First-order decay might occur for radionuclides as one isotope decays to another, or for biodegradation of organic contaminants by microbial-mediated processes. For a conservative tracer, $R = 1$ and $\lambda = 0$, and the ADE reduces to

$$\frac{\partial C}{\partial t} = -v\frac{\partial C}{\partial x} + D_L\frac{\partial^2 C}{\partial x^2} \tag{1b}$$

The longitudinal dispersion coefficient (D_L) is a measure of the spreading of a contaminant plume. The spreading is actually due to velocity variations at different scales and due to heterogeneities. Thus, D_L is typically considered to be the product of two terms, the dispersivity (α) and the average linear groundwater velocity, v.

$$D_L = \alpha_L v \tag{1c}$$

The dispersivity is generally thought to be a function of the length of the flow system (L) because large flow systems have more heterogeneity that causes more dispersion. Often, the following relation is used as a first estimate.

$$\alpha_L = 0.1\,L \tag{1d}$$

The retardation coefficient, R, is given by

$$R = 1 + \rho_b K_D/n \tag{2a}$$

$$K_D = K_{OC} f_{OC} \tag{2b}$$

where ρ_b is the dry bulk density of the soil (M/L^3), K_D is the soil water partition coefficient (L^3/M), for the contaminant, n is the porosity, K_{OC} is the organic carbon partitioning coefficient (L^3/M) for the contaminant and f_{OC} is the mass fraction of organic carbon in the soil (dimensionless). For a contaminant that sorbs to soil, $R > 1$ (except for some special cases).

Many solutions exist for this form of the ADE, expressed by Equation 1a, b, which can be found in many textbooks of hydrogeology or contaminant hydrogeology. We will choose a solution appropriate for the one-dimensional soil column illustrated in Figure 13.1.

For this soil column, we will pump in a slug input of dissolved tracer with concentration equal to C_0. The chemical will break through at the column outlet. The concentration at the column outlet will be less than the initial concentration because the slug spreads, or disperses, as it travels through the column.

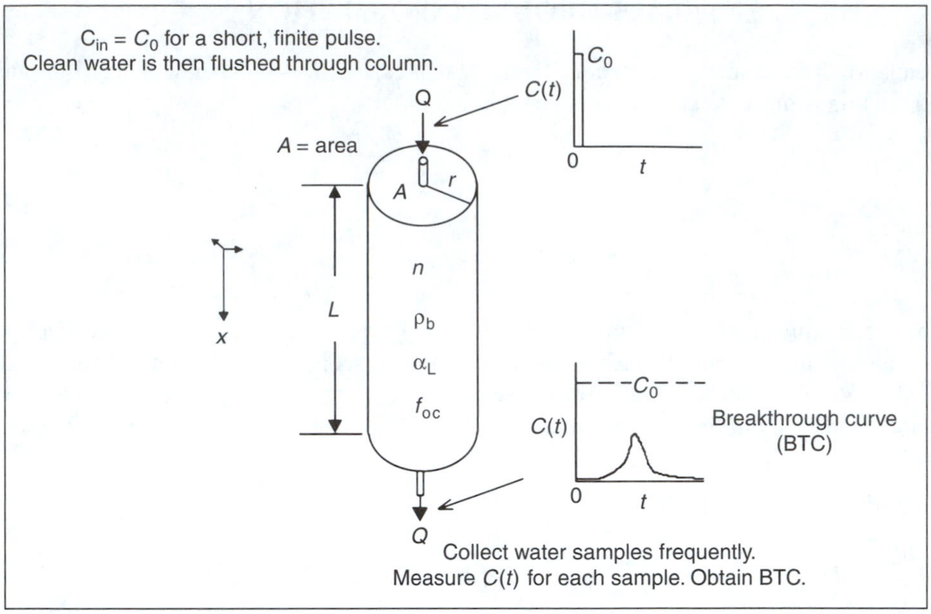

Figure 13.1—Soil column test.

For an instantaneous, or short pulse, input of tracer mass into a one-dimensional soil column with an initial concentration of zero, the following solution is provided (e.g., Bedient et al., 1999).

$$C(x,t) = \frac{m}{(4\pi Dt)^{1/2}} \exp\left[-\frac{(x - vt)^2}{4Dt} - \lambda t \right] \tag{3}$$

When concentrations are defined in the ADE as mass per volume of aqueous solution, the parameter m in the equation above is the aqueous-phase mass (m_{AQ}) in the transport domain per cross-sectional area available for flow (nA). For our column, x would be the length of the column (L). For a conservative tracer, the mass in the aqueous phase is the same as the chemical mass injected (m_{INJ}) into the flow domain because all chemical mass remains in the aqueous phase.

$$m = m_{AQ}/nA = m_{INJ}/nA \tag{4}$$

In other words, $m_{INJ} = C_0 T_0 Q$, where C_0 is the input concentration, Q is the volumetric flow rate of solution pumped into the column, and T_0 is the duration of the contaminant pulse or the length of time over which dissolved chemical is pumped into the column.

Equation 3 is valid for a chemical that decays, but does not sorb to aquifer material ($R = 1$).

For a sorbing chemical, such as TCE, a solution is readily found by inspection after rearranging Equation 1a. Specifically, if we divide Equation 1a by R, the following equation results.

$$\frac{\partial C}{\partial t} = -\frac{v}{R}\frac{\partial C}{\partial x} + \frac{D_L}{R}\frac{\partial^2 C}{\partial x^2} - \frac{\lambda}{R}C \tag{5}$$

The form of this equation is the same as Equation 1a. Thus, we can use the solution given in Equation 3 if we replace v with v/R, replace D with D/R, and replace λ by λ/R. We must also divide the M_{INJ} term by R to correct the total input mass to achieve the desired aqueous phase mass in the column (we are accounting for the mass sorbed to the soil phase).

$$C(x,t) = \frac{\dfrac{m}{R}}{\left(4\pi\dfrac{D_L}{R}t\right)^{1/2}}\exp\left[-\frac{\left(x - \dfrac{v}{R}t\right)^2}{4\dfrac{D_L}{R}t} - \frac{\lambda t}{R}\right] \tag{6a}$$

which simplifies to

$$C(x,t) = \frac{m}{(4D_L\pi R t)^{1/2}}\exp\left[-\frac{(Rx - vt)^2}{4D_L t} - \lambda t\right] \tag{6b}$$

where

$$m = m_{AQ}/nA = m_{INJ}/RnA \tag{7}$$

We will use these models to fit the outlet concentration versus time profile (or breakthrough curve). To fit the curve, we must vary the values for unknown parameters (e.g., R and λ) until the concentration versus time curve produced by the model (Equation 6b for a nonconservative tracer) matches the BTC to our satisfaction. Then we induce that these fitted values are representative of the soil in the column. This procedure works very well, but it is difficult to accomplish if there are too many parameters to fit. Thus, we strive to obtain estimates for as many parameters as possible using more direct methods. Thus, to calculate the dispersivity (and thus dispersion coefficient) for the soil in the column, we will use a method called the "temporal method of moments" to analyze the column effluent data independent of our analytical model. This method is described next.

Temporal Method of Moments

The method of moments is based on statistical theory. The zeroth temporal moment (M_0) is defined as the area under the C versus t curve, also called the breakthrough curve, or BTC.

$$M_0 = \int C\,dt = \Sigma C\Delta t \tag{8a}$$

For the BTC, numerous (C, t) data points make up the curve. One way to integrate under the BTC is to use the trapezoidal rule; for n (C, t) data points

$$M_0 = \sum_{k=1}^{n}\frac{C_k + C_{k-1}}{2}(t_k - t_{k-1}) \tag{8b}$$

where $k = 1$ is the first data point and $k = 0$ corresponds to $C = 0$ at $t = 0$.

The mass of chemical retrieved from the column effluent is equal to QM_0 for a constant flow rate through the column. If the flow rate varies with time, the following equation may be used.

$$\text{Mass recovered} = m_r = \sum_{k=1}^{n} \frac{C_k + C_{k-1}}{2}(t_k - t_{k-1})\left(\frac{Q_k + Q_{k-1}}{2}\right) \tag{9}$$

The first moment (M_1) is defined as

$$M_1 = \int Ctdt = \sum Ct\Delta t \tag{10a}$$

Again, using the trapezoidal method of numerical integration

$$M_1 = \sum_{k=1}^{n} \frac{C_k + C_{k-1}}{2}(t_k - t_{k-1})\left(\frac{t_k + t_{k-1}}{2}\right) \tag{10b}$$

Q is the volumetric flow rate, which is equal to the Darcy velocity (v_n) times the cross-sectional area (A) for flow (the column end plate area for a one-dimensional column test). M_1/M_0 gives the travel time of the center of mass of the solute plume. However, we are interested in the travel time of the tracer or of the front of the BTC. The travel time (T_i) of the tracer (where i refers to a particular chemical) is given by

$$T_i = M_1/M_0 - 0.5\,T_0 \tag{11}$$

The term T_0 is the input pulse length of the tracer, or the time that it takes to pump the tracer pulse into the input flow stream. The last term corrects for the time it takes the pulse center of mass to travel to the starting line.

The velocity (v) of the conservative tracer is simply

$$v = L/T_{\text{conservative}} \tag{12}$$

where L is the distance from the input to the observation point (the column length L for a laboratory test). This term v is also the average linear velocity of the water through the column and is the same v term used in the ADE and solutions earlier.

The contaminant velocity (v_c) is given by

$$v_c = L/T_{\text{contaminant}} \tag{13}$$

The retardation factor, R, is given by

$$R = \frac{T_{\text{contaminant}}}{T_{\text{conservative}}} = \frac{v}{v_c} \tag{14}$$

TCE, the contaminant of concern in this problem, may also degrade under certain groundwater geochemistry conditions (primarily, the dissolved oxygen in the water within the plume must be substantially depleted). This process should not slow down the plume beyond the retardation affected by sorption. However, the process will cause loss of TCE mass, and thus the concentrations within the BTC will be reduced.

PROCEDURES AND TASKS

Note: Throughout these tasks, be very careful that you use consistent units in each calculation!

Task 1—Review Test Procedures and Spreadsheet File

A vertical soil-column test using uncontaminated soil collected from the site has been completed. Two columns were used; in the first column, bromide was injected as a conservative tracer, along with dissolved TCE as a nonconservative (reactive) contaminant tracer. In this tracer test, you included a biological growth inhibitor to prevent biodegradation of the TCE; thus, sorption was the only process affecting TCE transport. In the second soil column, you again used bromide and dissolved TCE, but this time you did not use the inhibitor; thus, TCE transport was affected by both sorption and biodegradation. Analyze these three sets of data: bromide, TCE–sorption only, and TCE–sorption plus degradation.

The concentration input and flow parameters for the tests are given in the Excel spreadsheet LAB13.XLS on the CD included with this lab manual.

The results from the two tracer tests are essentially the same for bromide; thus, the results are averaged. The BTCs for bromide, TCE with sorption only, and TCE with both sorption and degradation, are plotted on two graphs in the spreadsheet to allow better evaluation of the data. Figure 13.2 shows the layout of the spreadsheet, including these two graphs.

The spreadsheet also contains an analytical model solution for each of the three tracer tests (Section 5, Figure 13.2). The results of the analytical solutions are also plotted on the graphs for comparison with the test data. Note that the model BTCs do not fit the BTCs plotted from the data, because the R, D, and λ parameters are not the proper ones for this soil test; the parameters currently used in these model solutions were arbitrarily selected. Your goal will be to determine the proper values of these parameters (R, D, λ) so the analytical model solutions match the observed data.

With the assistance of your lab instructor, look carefully at the spreadsheet to ensure that you understand the data and the figures presented. The worksheet labeled *Instructions* will help you understand the spreadsheet.

Task 2—Analyze Test Data to Determine Retardation Factor, Dispersion Coefficient, and Decay Constant

1. Use the method of moments described above to determine the travel times for both the conservative tracer (bromide) and the nonconservative, nondegrading tracer (TCE–sorption only). Use Section 6 of the spreadsheet for the calculation of M_0 and M_1.

2. Use this information to calculate a retardation factor for the TCE in the soil collected from the site. Note that you could calculate the velocity from the moment data using Equation 13. However, for a column test, this velocity (v) should be nearly equal to the known injection velocity (11 cm/day), which is the Darcy velocity or specific discharge (V) divided by porosity (n).

3. Calculate the mass (mg) of contaminant recovered from the column effluent for the two tracers. Compare the recovered mass to the total mass of chemical injected ($m_{\mathrm{INJ}} = C_0 T_0 Q_{\mathrm{INJ}}$, where C_0 is the concentration of the fluid injected into the column)).

   Should these numbers be exactly the same? Explain why or why not.

4. Calculate the M_0 and total mass (mg) recovered for the TCE tracer test where biodegradation was allowed. How much biodegradation occurred (percent of initial mass)?

   Is this amount significant? Why is it important to perform the calculation in No. 3 first?

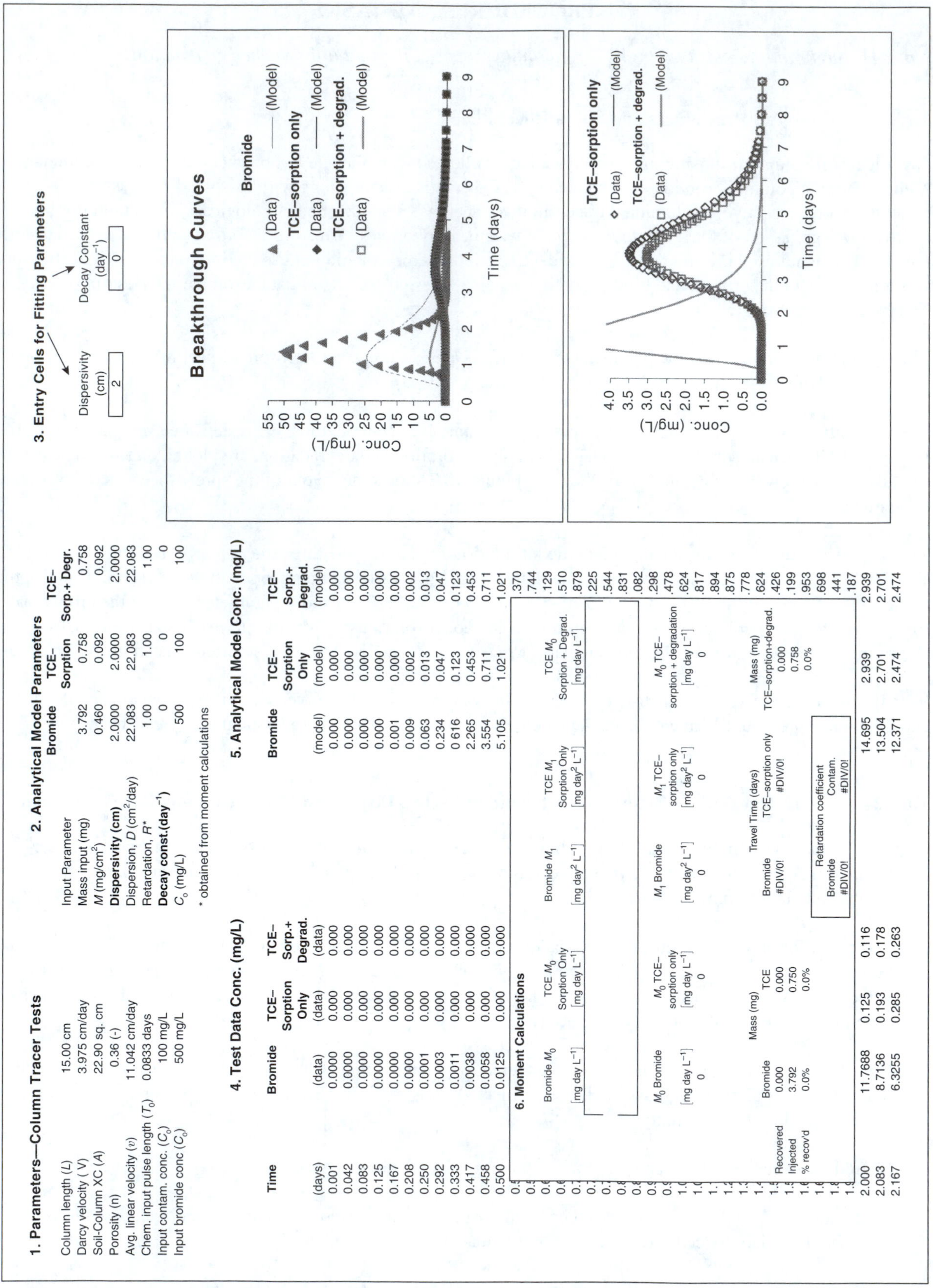

Figure 13.2—Organization of spreadsheet in file LAB13.xls.

5. In the spreadsheet, note Section 2, termed *Analytical Model Parameters*. These are the parameters required for use of Equation 6a, which is programmed in the spreadsheet in this section. Note that above the graphs, there are two cells labeled as fitting cells, one each for dispersivity (α_L) and the first-order degradation rate coefficient, or more simply, the decay constant (λ). These two parameters, along with the retardation factor (R), are the unknowns that you are trying to determine using the data from your column test. These fitting cell values will change the values in the *Analytical Model Parameters* section. The spreadsheet is set up this way so you can see the changes on the graph as you vary the parameters. In a sense, you are calibrating your one-dimensional contaminant transport model to the data from the column test. For R, you should enter a value of 1.0 for the conservative tracer (not retarded); for both contaminant (TCE) tracer tests, you should enter the value of R that you calculated in No. 2.

   Calculate a reasonable first estimate of the longitudinal dispersivity (α_L) for the soil column.

6. In this next procedure, you will be fitting α for the contaminant data sets and λ for the biodegrading contaminant data set. For dispersivity, enter the first estimate calculated above, and then vary the value in the fitting cell while observing the model curve change in the graph until the bromide model curve fits the bromide data.

   What is your best estimate for dispersivity?

   Is it larger or smaller than your first estimate? Discuss some reasons for any difference.
   The model will calculate the dispersion coefficient, D, from your final value of dispersivity, α, and groundwater velocity, v.

7. You will use the dispersivity value you determined using bromide for both the contaminant (TCE) tests because dispersivity is a function of the porous medium, and you used the same soil in both of the columns. Thus, you should use the fitted α value for both sets of contaminant data. Your bromide and TCE (sorption only) model curves should fit the data very well.

   Now vary the decay constant until the TCE (sorption + degradation) model fits the data from the degrading-TCE test.

   What is your best estimate for the decay constant, that is, the first-order degradation rate coefficient?

Task 3—Evaluate and Apply Results

1. It might be argued that one could try to fit all the model parameters from the start, without first calculating the dispersivities using the method of moments. Do you think this is a good idea? Why, or why not?

2. The transport of contaminants in the field is actually three-dimensional. Often, field plumes are modeled using one-dimensional analytical solutions. What are some obvious problems with this approach?

3. Estimate the time for the groundwater to travel from the spill location to the stream.

4. Estimate a field-scale value for the longitudinal dispersivity (α_L) in meters, and then calculate a field-scale value for the dispersion coefficient (D_L) (m^2/day). Assume that the field scale of interest is 50 m.

5. We are often concerned about the maximum concentration that can occur along the plume centerline. Based on the particular problem described for this laboratory, use of a two-dimensional ADE may be valid. For a two-dimensional problem, a value for the transverse dispersion coefficient (D_T) would also be required. It often is assumed that $D_T = 0.1\, D_L$ (or $\alpha_T = 0.1\, \alpha_L$). Using this approximation and assuming the TCE does not degrade, calculate the maximum TCE concentration along the centerline of plume travel at the time the groundwater from the spill site reaches the stream, using

$$C_{max} = \frac{C_0 A}{4\pi t \sqrt{D_T D_L}} \tag{16}$$

where A is the surface area over which the spill occurred and C_0 is the concentration of the contaminant in the spilled liquid.

How does this value compare to the EPA-regulated maximum contaminant level for TCE (0.005 mg/L)?

Estimate the location for this maximum.

6. For a degrading contaminant, such as the TCE you are dealing with, an estimate of C_{max} may be obtained by multiplying Equation 16 by $e^{-\lambda t}$.

Calculate the maximum concentration of TCE along the centerline of plume travel at the time the plume reaches the stream, using the decay constant you calculated from the laboratory test.

Will there be a problem when the plume reaches the stream? What is your bottom-line assessment of the problem?

7. Your goal in conducting these tests was to estimate some input parameters to model the contaminant transport at the site described above. What are some limitations to using each of the parameters measured with soil-column tests for predicting field transport of the plume?

What recommendations could you provide to reduce the uncertainty associated with using lab-measured parameters for field studies?

LAB 14

CONTAMINANT TRANSPORT II: MULTIDIMENSIONAL TRACER-TEST ANALYSIS AND CONTAMINANT-PLUME MODELING

PURPOSE: Become familiar with the use of field tracer tests in modeling a contaminant plume.

OBJECTIVES: Determine input parameters for a mathematical contaminant transport model, using data from a field tracer test.

> Use the spatial method-of-moments to analyze spatial monitoring-well data to estimate average linear groundwater velocity and dispersion coefficients.
>
> Estimate retardation factors from field data.
>
> Determine if degradation occurs, and choose an appropriate biological decay constant.

Use these input parameters in a multidimensional analytical solution of the advection-dispersion equation (ADE) to assess the potential for pollution from a carbon tetrachloride plume.

OVERVIEW: Transport of contaminants in groundwater is modeled to determine transport velocity and contaminant-plume shape. The models used for this, including simpler analytical models and more complex numerical models, require input parameters that characterize both the aquifer (hydraulic gradient, hydraulic conductivity, porosity, dispersivity, and heterogeneity) and the contaminant (degradation and retardation). Modeling success, of course, depends on the accuracy of estimating these input parameters. Three approaches may be used: The input parameters may be estimated from a general knowledge of the aquifer (the least desirable approach), the parameters may be estimated from laboratory analyses of aquifer samples, or the parameters may be estimated from field tests in the aquifer itself, (e.g. using tracers).

Labs 13, 14, and 15 labs deal with contaminant transport.

> Lab 13 deals with the estimation of input parameters from lab analyses of aquifer samples.
>
> Lab 14 estimates the parameters by analysis of a tracer injected into the aquifer and uses the derived parameters in a simple analytical model to estimate travel of a contaminant plume.
>
> Lab 15 models a contaminant plume and investigates the difference between assuming aquifer homogeneity, using a lumped-dispersion-coefficient approach, and modeling aquifer heterogeneity.

REQUIRED MATERIALS

PC computer

Spreadsheet software is preferred (e.g., Microsoft Excel), but a calculator and graph paper can be used.

INTRODUCTION

Problem Statement

As an environmental consultant specializing in groundwater contamination, you are concerned with a spill of carbon tetrachloride (CT) discovered on your client's property. CT is a chlorinated organic solvent that is thought to cause cancer in humans. For reasons of human health, social implications, and politics, it is critical that the plume of dissolved CT leaving the spill site be prevented from reaching a point 39 m away.

Your client wants to aggressively pursue characterization of the plume, and, if necessary, cleanup. Because of the severity of the problem, your firm has been authorized to install numerous wells, both to monitor the CT plume and to conduct a tracer test. The tracer test will provide information about the aquifer that you can then use to model the transport of the dissolved CT plume.

In its pure form, CT is an organic liquid that is denser than water and is immiscible with water. Thus, CT is a dense, nonaqueous-phase liquid, or DNAPL, whose presence at the site has been verified by soil cores. CT concentrations in soil within the source zone exceed 50,000 mg/kg. The DNAPL is likely to be relatively immobile at the site, but the CT will dissolve very slowly into the groundwater, and it will likely act as a source of groundwater contamination for decades.

Based on company records, you deduce that the spill occurred about seven years ago. Tests found relatively high, constant aqueous concentrations of CT (150 mg/L) within a "hot spot" of the aquifer corresponding to the DNAPL-contaminated soil. DNAPL is present throughout the entire thickness of the saturated zone (5 m $\pm$ 0.3 m throughout the site) within the sandy aquifer. The dimensions of the spill in the saturated zone are about 2 m perpendicular to groundwater flow and 1.5 m in the direction of groundwater flow.

Monitoring wells were installed near the spill site, and the monitoring-well network was expanded downstream to monitor the plume. These monitoring wells were also used to conduct a tracer test in which bromide, a conservative tracer (a solute that does not react with the porous media), was injected in one of the wells and its travel with the groundwater was monitored. Analysis of the data from this tracer test will provide you with input parameters for a mathematical model that you will use to estimate plume transport. Some of the model input is available from laboratory analyses of soil cores from the site.

Field tracer tests are expensive, but not uncommon at important sites. One hopes that the expense of conducting these tests will be mitigated by the long-term savings gained from a better understanding and more accurate prediction of contaminant flow and transport at the site.

BACKGROUND INFORMATION

Theory of Chemical Transport in Groundwater

The most common mathematical governing equation derived to model the physics of chemical transport in groundwater is the ADE. The three-dimensional form of the ADE, including instantaneous reversible partitioning of a contaminant to another phase (e.g., sorption to soil) and linear, first-order degradation (e.g., radioactive decay or simple biodegradation) is given by the partial differential equation (PDE).

$$R\frac{\partial C}{\partial t} = -\frac{\partial (vC)}{\partial x} - \frac{\partial (vC)}{\partial y} - \frac{\partial (vC)}{\partial z} + \frac{\partial}{\partial x}\left(D_X\frac{\partial C}{\partial x}\right) + \frac{\partial}{\partial y}\left(D_Y\frac{\partial C}{\partial y}\right) + \frac{\partial}{\partial z}\left(D_Z\frac{\partial C}{\partial z}\right) - \lambda C \qquad (1)$$

This equation is rigorous for contaminant-plume travel parallel to one of the major axes (typically the X-axis). C is the concentration of the contaminant in water [M/L^3, typically mg/L]. R is the retardation factor (to be discussed in more detail later), typically used to represent "retardation" of contaminant velocities with respect to the groundwater velocity due to sorption and desorption of the contaminant to the soil. R is dimensionless. $v(L/T)$, is the average linear groundwater velocity along the direction of travel of the plume, typically the x-direction.

D_X, D_Y, and D_Z are the dispersion coefficients in the *x*-, *y*-, and *z*-directions (L^2/T).

λ is a first-order degradation rate coefficient (T^{-1}), usually reported using the units (day^{-1}).

Typically, this equation is simplified to either one or two dimensions, and assumptions regarding spatial homogeneity are invoked before solving. For example, velocities and dispersion coefficients are typically assumed to be spatially independent, and thus are moved outside the differential. With these assumptions, the two-dimensional (*X–Y*) ADE reduces to

$$R\frac{\partial C}{\partial t} = -v\frac{\partial C}{\partial x} - v\frac{\partial C}{\partial y} + D_L\frac{\partial^2 C}{\partial x^2} + D_T\frac{\partial^2 C}{\partial y^2} - \lambda C \qquad (2)$$

where $D_L = D_X$, typically called the longitudinal dispersion coefficient (along the direction of flow), and $D_T = D_Y$, called the transverse dispersion coefficient. This equation is valid for cases where the contaminant is uniformly spread throughout the thickness of the aquifer or, in a more practical sense, when the data are obtained from wells that are screened over the entire thickness of the saturated zone or over some zone that is thicker than the contaminant plume. The concentration variations with depth are averaged in the well bore so that resolution in the *Z*-direction is lost.

Dispersion is a measure of the spreading of a contaminant in the subsurface because of various processes. A good discussion on factors causing contaminant-plume dispersion is included in most texts on contaminant hydrogeology (e.g., Domenico and Schwartz, 1998; Fetter, 1999; Bedient et al., 1999). These processes include aqueous-phase molecular diffusion, pore-scale spreading due to velocity variations within the pore, and spreading at various scales because of velocity variations that are caused by porous-media heterogeneities at various scales. For example, at the medium and larger scale, low-permeability lenses or layers will cause the plume to spread around the layers. Higher velocities will cause increased spreading. Dispersion because of velocity variations is often termed "mechanical dispersion."

Mechanical dispersion typically overwhelms dispersion caused by of molecular diffusion for field scenarios (except for very small velocities). Thus, molecular diffusion is often neglected. Field-dispersion coefficients are thus dependent on the groundwater velocity and on the scale of the plume. Dispersion coefficients measured in laboratory soil columns are typically much smaller than those measured in the field, even when the soil sample is collected from the field site of interest.

The following expression is typically used for dispersion coefficients for field problems:

$$D_L = \alpha_L v \qquad D_T = \alpha_T v \qquad (3a)$$

where α_L is the longitudinal *dispersivity* [*L*] and α_T is the transverse *dispersivity*. The parameter v is the groundwater velocity. The longitudinal dispersivity is typically much larger (e.g., a factor of 10 or more) than the transverse dispersivity. The dispersivity is a *lumped parameter* that accounts for the impact of porous media characteristics on contaminant spreading. It is also a function of scale. Many correlations between dispersivity and scale have been developed in the literature. A simple expression is:

$$\alpha_L = 0.1\,\text{L} \qquad (3b)$$

where *L* is the scale of the transport problem or experiment (e.g., the length of the plume in a field situation).

Many models have been developed to simulate contaminant fate and transport in porous and fractured media. These may generally be categorized into numerical models and analytical models. Numerical models are algebraic discretizations of the PDE governing groundwater flow and contaminant transport (e.g., the ADE). Some numerical models solve the groundwater flow equation to solve for velocities in the spatial domain of interest and use these velocity values in the ADE to obtain solutions for contaminant transport. Analytical models are exact solutions to these partial differential equations. To obtain solutions analytically, many simplifying assumptions must be made. However, these solutions are generally much easier to use than more complex numerical models. For example, in this laboratory, you will code analytical solutions to equation 2 for the case of a *conservative* chemical (no retardation or degradation) in an Excel spreadsheet.

A solution to Equation 2, for conditions of an instantaneous (slug) input of a conservative chemical over a surface area, A, that can be assumed initially to penetrate the entire monitored depth of the aquifer, follows (after de Josselin de Jong, 1958)

$$C(x, y, t) = \frac{C_0 A}{4\pi (D_L D_T)^{1/2}} \exp\left[\frac{((x - x_0) - vt)^2}{4D_L t} - \frac{(y - y_0)^2}{4D_T t}\right] \tag{4}$$

where (x_0, y_0) is the location of the center of the spill, (x, y) is the point of interest or location of monitoring well, and $v = v_x$ = average linear groundwater velocity along the travel path. If first-order degradation is to be included, then the term $-\lambda t$ would be added inside the brackets in Equation 4.

The retardation factor R is defined as follows:

$$R = 1 + \rho_b K_D / n = v / v_c \tag{5a}$$

$$K_D = K_{OC} f_{OC} \tag{5b}$$

where ρ_b is the dry bulk density of the soil (M/L^3), K_D is the soil water partition coefficient (L^3/M) for the contaminant, n is porosity, K_{OC} is the organic carbon partitioning coefficient (L^3/M) for the contaminant, and f_{OC} is the mass fraction of organic carbon in the soil (dimensionless). For an organic contaminant that sorbs to soil, $R > 1$ generally. The term v_C is the contaminant velocity, and v is the groundwater velocity (also the velocity of a conservative, non-reactive, tracer). To include contaminant retardation (R) in Equation 4, the terms D and v should be replaced by the terms D/R and v/R. If degradation and retardation are included, degradation would be added as described in the paragraph above, except $-\lambda t/R$ would be included in the bracketed term of Equation 4.

A three-dimensional analytical solution for the ADE with the condition of a constant, continuous source is given here. This solution includes biodegradation and sorption (retardation). Note that for this equation, the term v_c *is* the velocity of the contaminant or tracer, not necessarily the velocity of the groundwater.

$$C(x, y, z, t) = \frac{C_0}{8} \exp\left(\frac{x}{\alpha_X 2}\left(1 - \left(1 + 4\lambda\frac{\alpha_X}{v_c}\right)^{1/2}\right)\right) erfc\left(\frac{x - v_c t\left(1 + 4\lambda\frac{\alpha_X}{v_c}\right)^{1/2}}{2\sqrt{\alpha_X v_c t}}\right)$$
$$\left[erf\left(\frac{y + Y/2}{2\sqrt{\alpha_Y x}}\right) - erf\left(\frac{y - Y/2}{2\sqrt{\alpha_Y x}}\right)\right]\left[erf\left(\frac{z + Z}{2\sqrt{\alpha_Z x}}\right) - erf\left(\frac{z - Z}{2\sqrt{\alpha_Z x}}\right)\right] \tag{6}$$

For this model, Y is the width of the source zone (perpendicular to the direction of groundwater flow), and Z is the depth of the source zone below the water table. In our problem, the DNAPL is the contaminant source. The origin of the coordinate system for this solution is assumed to be at the center of the source zone horizontally and at the water table vertically (i.e., $x = 0$ and $y = 0$ in the lateral center of the source zone, $z = 0$ at the water table). The α terms are the dispersivities in the x-, y-, and z-directions. Often, it is assumed that $\alpha_Z = \alpha_Y = \alpha_T$. The term α_X corresponds to the longitudinal dispersivity, α_L.

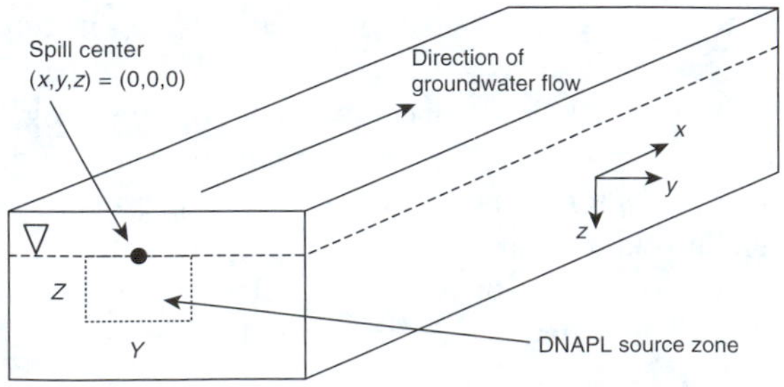

The error function (erf) and complementary error function ($erfc$) used in Equation 6 are tabulated in several contaminant hydrology books. The two functions are related by

$$erfc(b) = 1 - erf(b) \tag{7a}$$

Many spreadsheets are also able to calculate these functions. However, most spreadsheets cannot compute *erfc* and *erf* of a negative number. For these cases, the following equations should be useful.

$$erf(-b) = -erf(b) \quad \text{and} \quad erfc(-b) = 1 + erf(b) \tag{7b}$$

If your spreadsheet software cannot evaluate these functions, you can use

$$erf(b) \approx \sqrt{1 - \exp\left(\frac{-4b^2}{\pi}\right)} \tag{7c}$$

Note that the exponential function in (7c) is appropriate only for $b > 0$. For $-b$, use $erf(-b) = -erf(b)$.

Determination of Aquifer Contaminant Transport Parameters

To use a *relatively* simple analytical model like those expressed by Equations 4 and 6, or a more complex numerical model, one must determine the transport model input parameters (C_O, D, α_T, α_L, v, R, λ), generally from the results of field or laboratory data. In some cases, data collected at similar sites, which may be found in the literature, are used.

The C_O value for CT is assumed to be the persistent concentration ($\sim$150 mg/L) in the center of the DNAPL source zone (determined from contour plots of CT concentrations measured from groundwater samples). Retardation and the decay constant are also estimated from field data, as will be described later. The primary focus of this lab is to use data from a conservative tracer test to estimate field values for groundwater velocity and for the dispersivities (from which the dispersion coefficient can be calculated). The tracer test is termed a natural-gradient test because we will not use pumps but will rely on the natural hydraulic gradient to transport the tracer.

For this tracer test, a slug of dissolved bromide (as potassium bromide, or KBr) is injected into a well near the center of the DNAPL source zone. The bromide is a conservative tracer, meaning it will not react with the subsurface solids, and will travel at the same rate as the groundwater. The bromide will move due to advection and dispersion, and a bromide plume will be created. The plume will not have a constant source, as does the CT plume; rather the plume will behave as illustrated in the bird's-eye view schematic shown here.

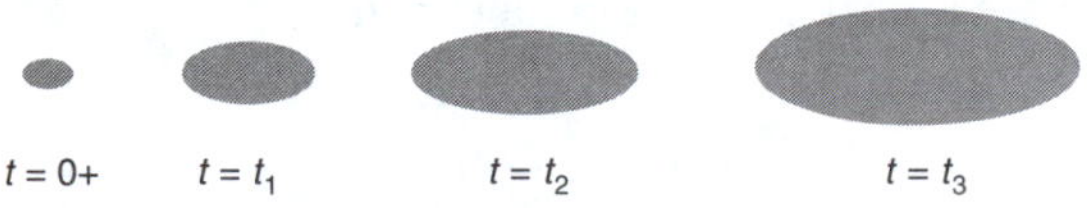

The CT plume can be conceptualized as shown (where a lighter shade represents the plume at a later time).

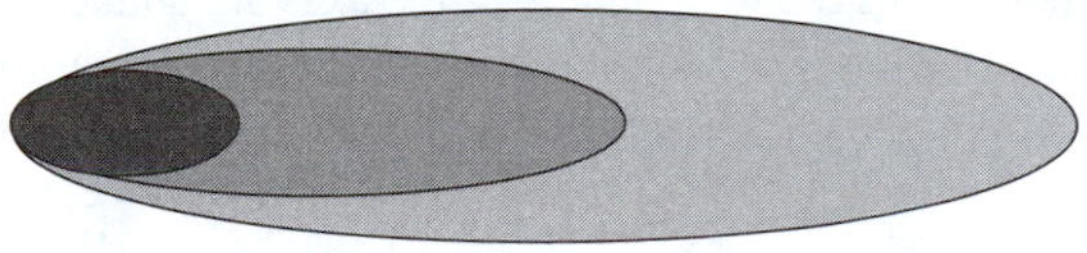

Of course, the plume shapes observed in the field are not as elliptical as shown here. For the tracer test, groundwater samples from each monitoring well are collected over time and bromide concentrations for each sample are determined. These bromide concentration data will be used to estimate dispersion and groundwater velocity.

The final monitoring-well network is illustrated below. Note that the higher density of monitoring wells is aligned along the direction of plume transport. Of course, all of these wells were not drilled at one time. The wells were drilled to stay ahead of the bromide plume over the course of a year. The existing CT plume also provided guidance on where wells should be located. The monitoring-well grid contains 73 monitoring wells screened over the entire saturated thickness. The grid is not uniform for practical reasons (damaged wells, etc.).

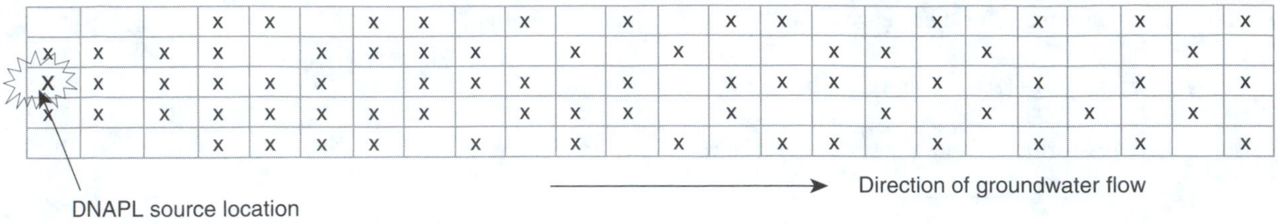

For the test, a slug of KBr solution (i.e., KBr dissolved in water) was injected into well X (shown above) in the center of the DNAPL source zone. A total of 155 g of bromide was used in the injected slug.

To determine values for dispersivity, velocity, and flow direction from bromide concentration data, we will use the so-called spatial method of moments (Freyberg, 1986). It has been proposed that this method works better if regularly spaced data are used. Because wells are usually not regularly spaced (as in our case), interpolation methods are often used to "regularize" the data to a uniform spacing prior to using the method.

The zeroth spatial moment (M_0) indicates the total amount of chemical tracer mass in the dissolved phase. For fully screened monitoring wells, M_0 is expressed by

$$M_0 = \sum_{i=1}^{N} b_i A_i C_i n_i(x, y) \quad i = 1, N \quad N = \text{number of concentration values} \tag{8}$$

where b_i is the thickness of the aquifer at the sampling well, A_i is the surface area (or weight) associated with each well, and n_i is porosity at each sampling location. Often, a constant porosity and constant thickness are assumed, and the n and b terms are moved outside the summation. If the monitoring wells are not fully screened, but are screened deeper than the plume, the parameter b is often set to the screened-interval lengths. For these cases, it is important that the screens be long enough to sample the entire plume. The A_i for each well can be calculated using a variety of methods. For example, one of the methods used to assign the watershed subareas to each rain gage in a watershed could also be used for this purpose.

We will define the direction of plume movement as the x-direction. For a slug input of chemical, the x-coordinate of the mean position of the plume (x_C, or the position of the center of mass) is given by the *normalized* first moment in the x-direction ($\overline{M_{1X}}$) minus the x-coordinate of the spill location (x_0).

$$x_C = \overline{M_{1X}} - x_0 = \frac{M_{1X}}{M_0} - x_0 = \frac{\sum_{i=1}^{N} b_i A_i C_i n_i x_i}{M_0} - x_0 \tag{9}$$

Note that M_{1X} is the first spatial moment. x_i is the x-coordinate of each well. The y_C position can be determined from a similar equation, where y_i and y_0 are used instead of x_i and x_0. If the axes are aligned so that the x-axis is along the travel path of the plume (as is the case for this laboratory exercise), then $y_c = 0$ and $(x_C - x_0)$ gives the distance (d_C) traveled by the plume. If several synoptic sampling rounds are conducted, then various values of d_C can be plotted versus time. The slope of this plot is the mean velocity of the contaminant plume over the distance from x_0 to x_c.

$$v_x = v_{\text{bromide}} = v = dx_C/dt \tag{10}$$

For a conservative tracer like bromide, this velocity should also be the mean velocity of the groundwater. For a non-conservative tracer (e.g., one that sorbs to soil, like most organic contaminants), the velocity of the plume will generally be less than that of the groundwater.

To determine the dispersivity, we need to calculate the variance, S, which is based on the values of the second spatial moments M_{2x} and M_{2y}):

$$S_x = S_L = \overline{M_{2x}} - x_C^2 = \frac{M_2}{M_0} - x_C^2 = \frac{\sum_{i=1}^{N} b_i A_i C_i n_i x_i^2}{M_0} - x_C^2 \tag{11a}$$

$$S_y = S_T = \overline{M_{2y}} - y_C^2 = \frac{M_2}{M_0} - y_C^2 = \frac{\sum_{i=1}^{N} b_i A_i C_i n_i y_i^2}{M_0} - y_C^2 \tag{11b}$$

These equations are valid when the x-direction is aligned along the direction of plume transport.
To calculate the longitudinal dispersion coefficient, plots of S_L versus sampling time are constructed. The slope of the line is equal to $2D_L$ with the S_L intercept equal to zero, or

$$S_L = (2D_L)t \tag{12}$$

PROCEDURE

You want to characterize the site using a field tracer test and use the information to develop an analytical model to predict flow of CT.

Task 1—Analyze Tracer Test Data

1. Write the dimensions of each term in the ADE (use M for mass, L for length, T for time).

2. Write consistent units for M_0, M_1, M_2, X_C, Y_C, S_L, and S_T into the appropriate brackets for Tables 14.1a and 14.1b (use actual units; e.g., length should be in ft, m, or cm, etc.)

3. Perform dimensional analysis, and verify that each of Equations 6–12 is dimensionally consistent.

4. The locations of the interpolated data points and the associated interpolated bromide concentration data are on the CD included with this laboratory manual in an Excel spreadsheet file Lab14.xls. Also read the *Instructions* worksheet in the spreadsheet. Look carefully at the appropriate data file with your laboratory instructor, and ensure that you understand the setup. Note that the monitoring-well network has been expanded after each sampling round in an attempt to monitor the entire plume as it spreads. Recall that the data entries are not all from actual monitoring wells; the actual number of monitoring wells is equal to about one-third the number listed.

5. Calculate the moments, percent bromide recovered in the monitoring wells, center coordinates of the plume, and variances for each sampling time (preferably using the spreadsheet provided). Write the results in Tables 14.1a and 14.1b.

Note: Before using the spreadsheet provided, carefully read the instructions provided on the "Instructions" worksheet in the spreadsheet.

6. Ideally, you want to account for 100 percent of the injected tracer mass with your sampling system. This is difficult in the field, and most hydrogeologists agree that a 90% recovery rate is acceptable. Based on this rule of thumb, which of the above sampling rounds would you consider to have unacceptable error?

7. List probable reasons for any significant discrepancies and potential actions to correct the problem.

Table 14.1a						
Time (days)	**(M_0) (Mass) ( )**	**Percent of Injected Mass Recovered**	**(M_{1X}) ( )**	**(M_{1Y}) ( )**	**(M_{2X}) ( )**	**(M_{2Y}) ( )**
0						
90						
180						
270						
360						

Note: M_0 is equal to the total mass of tracer in the plume.

Table 14.1b				
Time (days)	X_C ( )	Y_C ( )	S_L ( )	S_T ( )
0				
90				
180				
270				
360				

8. List some general problems that might impart error to this field tracer-test analysis.

9. Plot X_C versus time. From this plot, determine the slope and calculate the average linear velocity of the groundwater (m/day). Enter the result in Table 14.2 on page 111. What is the average velocity in the Y-direction? Why? What does the slope of the line tell you about the longitudinal velocity of the plume over time? What would the plot look like if the plume slowed down over time? For parameters in Table 14.2, list the source of the data (e.g., field tracer test, laboratory analysis of field-collected soil samples, from the literature, etc.).

10. Plot the x-variance and y-variance versus time (two different plots), and determine the slope. Calculate the longitudinal and transverse dispersion coefficients and enter the results in Table 14.2. Also calculate and enter the dispersivities in Table 14.2.

Task 2—Estimate Retardation Coefficient and Degradation Rate Constant

1. Your company's soil physicist estimated an average organic carbon fraction (f_{OC}) at the site of 0.005 ± 0.001, based on analysis of several hundred soil samples collected during well drilling and probing. The average bulk density was 1.85 ± 0.06 kg/L, and the porosity was estimated to be 0.31 ± 0.02. Bedient and colleagues (1999) list a K_{OC} value of 217 for CT. Estimate a retardation factor based on this information, and enter the value in Table 14.2.

2. Calculate an average linear *contaminant* velocity, v_c. Enter the result in Table 14.2.

3. Your company's microbiologist believes that, based on site geochemistry, carbon tetrachloride should be degraded in the field by a process known as reductive dechlorination. Your laboratory technicians, however, cannot reproduce degrading conditions in the laboratory (lab-measured degradation rates are equal to 0). Your

hydrologic modeling expert consultant contends, however, using an analytical model developed for the site, that degradation is actually occurring at a rate of 0.004 day^{-1}. Add the value you will use to Table 14.2.

Table 14.2—COMPILATION OF MODEL-INPUT PARAMETERS

Parameter	Value	Units	Source of Data and Equation Used
C_0 (CT)			
Y			
Z			
v			
f_{OC}			
K_{OC}			
n			
ρ_b			
R			
v_c			
D_L			
D_T			
α_X			
α_Y			
α_Z			
λ			

Task 3—Estimate the Transport of the CT Plume

1. Based on the contaminant velocity calculated earlier, how far from the source zone would you expect the contaminant to move in 5 years? 10 years?

2. Equation 6 is the more appropriate of the two analytical solutions described for modeling the transport of the CT plume. Explain why. Which is more appropriate for modeling a bromide plume? Explain why.

3. You want to determine which degradation rate is best to use in the modeling analysis. The lab measured a value of zero. The modeling expert recommended a rate of 0.004 day^{-1}. You can do this in a simple way by comparing model results for CT concentrations to the actual concentrations measured in some wells recently, assuming the spill occurred 7 years ago (as company records suggest; recall the introduction to this laboratory exercise). For this exercise, compare the results of model-simulated concentrations using the two degradation rates (0 day^{-1} and 0.004 day^{-1}) to the concentrations collected in six monitoring wells (Table 14.3) and enter the results. The file Lab14.xls contains a spreadsheet model for Equation 6 on worksheet *CT Plume Model*. The lab instructor will explain this spreadsheet model to you. You should use the columns with the XLS functions *erf* and *erfc*, if available. If not, you can use the columns with the exponential approximation for *erf*. In both cases, be careful to use the appropriate equations (from Equations 7a and 7b) when the argument of either *erfc* (b) or *erf* (b) is negative (e.g., $b < 0$).
   (a) Which degradation rate do you think is more reasonable?
   (b) Do you think that data from six wells is enough to distinguish between the two rates?
   (c) Do you think data from six wells is sufficient to determine the exact degradation rate?

Table 14.3—CONCENTRATIONS IN SIX MONITORING WELLS			
MW Coordinates (x, y)	**Measured C (mg/L)**	**Model C (mg/L) $\lambda = 0.004\, d^{-1}$**	**Model C (mg/L) $\lambda = 0\, d^{-1}$**
(5,0)	42.0		
(7,0)	17.8		
(9,0)	12.1		
(12,0)	4.90		
(27,0)	0.112		
(39,0)	0.002		

4. The receptor point of concern is a monitoring well, located about 39 m downgradient of the spill site. If the spill coordinates are $(x,y,z) = (0,0,0)$ (m) and the well screen coordinates are $(39,0,0)$, calculate the concentrations at the monitoring well at two-year intervals from 1 to 21 years, using the degradation rate you feel is most appropriate, and enter in Table 14.4. Based on this model, describe the trend exhibited in Table 14.4. Based on the model, will the maximum allowed concentration of CT (0.005 mg/L) be exceeded at this point? If so, when?

Table 14.4	
Time (yrs)	**Concentration (mg/L)**
1	
3	
5	
7	
9	
11	
13	
15	
17	
19	
21	

Task 4—Conclusions

1. Would you feel comfortable using the results of the analytical modeling for final predictions of transport times and concentrations of the CT plume? Why, or why not?

2. Discuss some of the difficulties in using models for prediction of contaminant transport.

3. Describe some of the benefits of using models for prediction of contaminant transport.

4. Considering your answers to the previous questions, what future actions would you recommend at the site, based on these model results?

LAB 15

CONTAMINANT TRANSPORT III: MODELING GROUNDWATER FLOW AND CONTAMINANT PARTICLEFLOW

PURPOSE: Investigate dispersion of contaminant particles in homogeneous and heterogeneous porous media through use of a numerical model.

OBJECTIVES: Understand the cause of contaminant dispersion in porous media.

Appreciate the limitations and conveniences of using the lumped-dispersion-coefficient approach in modeling contaminant-plume transport.

Become familiar with the use of numerical models for simulating contaminant-plume movement.

Visualize groundwater flow in two dimensions.

REQUIRED MATERIALS: PC computer, calculator

OVERVIEW: Transport of contaminants in groundwater is modeled to determine transport velocity and contaminant-plume shape. The models used for this, including the numerical model used in this lab, require input parameters that characterize both the aquifer (hydraulic gradient, hydraulic conductivity, porosity, dispersivity, and heterogeneity) and the contaminant (degradation and retardation). Modeling success, of course, depends on the accuracy of estimating these input parameters. Three approaches may be used: The input parameters may be estimated from a general knowledge of the aquifer (the least desirable approach), the parameters may be estimated from laboratory analyses of aquifer samples, or the parameters may be estimated from field tests in the aquifer itself, (e.g. using tracers).

Labs 13, 14, and 15 deal with contaminant transport.

> Lab 13 deals with the estimation of input parameters from lab analyses of aquifer samples.
>
> Lab 14 estimates the parameters by analysis of a tracer injected into the aquifer and uses the derived parameters in a simple analytical model to estimate travel of a contaminant plume.
>
> Lab 15 models a contaminant plume and investigates the difference between assuming aquifer homogeneity, using a lumped-dispersion-coefficient approach, and modeling aquifer heterogeneity.

INTRODUCTION

As discussed in most hydrogeology textbooks, the spreading, or dispersion, of solutes or contaminants in an aquifer is due to spatial variability in the velocity field. The velocity variations at field scales are due to heterogeneities in the aquifer. The use of a dispersion coefficient is a simple way to account for these velocity variations, but we use the dispersion coefficient only because we cannot usually characterize the heterogeneities in sufficient detail to accurately model plume velocities at all locations in the aquifer. If we could account for all the heterogeneity, a dispersion coefficient would not be necessary. We could use an advection-only groundwater flow model and assume the contaminants are particles that flow along the same flow paths as the groundwater (although not necessarily at the same velocity). If molecular diffusion (spreading due to concentration gradient) is important, we would also need to include this process, but it is usually negligible at typical groundwater flow velocities (exceptions: very low permeability media, very low hydraulic gradients).

Because we cannot typically characterize aquifer heterogeneity in sufficient detail, we often use an ADE (see Labs 13 and 14) to model contaminant movement. The ADE requires use of a lumped parameter, called the dispersion coefficient (D), to simulate spreading. For a two-dimensional model simulation, we have spreading in the horizontal and vertical directions; thus a horizontal dispersion coefficient (D_x) and a vertical dispersion coefficient (D_y) are required.

$$D_x = \alpha_x v_x \quad [L^2/T] \tag{1a}$$

$$D_y = \alpha_y v_x \quad [L^2/T] \tag{1b}$$

The terms α_x, α_y, are the horizontal and vertical "dispersivities." These terms have dimensions of length $[L]$ and are properties of the porous media. As heterogeneity increases, α also increases. These coefficients are multiplied by the horizontal flow velocity to simulate the increase in spreading due to increases in plume velocity. This approach is valid when the dispersion of contaminants in a flow field is Fickian—that is, the dispersion of contaminants resembles uniform diffusion around the center of mass of a contaminant plume.

For flow of particles in a field of spatially nonuniform velocities (e.g., a heterogeneous domain), we can define the spatial variances of the particle plume in the x- and y-directions (S_x and S_y, respectively) as the spatial variance of particle positions

$$S_x = \frac{1}{N}\sum_{i=0}^{N}(x_i - x_c)^2 \quad \text{and} \quad S_y = \frac{1}{N}\sum_{i=0}^{N}(y_i - y_c)^2 \tag{2}$$

where N is the total number of particles, x_i and y_i are the x- and y-coordinates of the ith particle, and x_c and y_c denote the x and y positions of the center of mass of the particle plume, defined as

$$x_c = \frac{1}{N}\sum_{i=0}^{N} x_i \quad \text{and} \quad y_c = \frac{1}{N}\sum_{i=0}^{N} y_i \tag{3}$$

The variance values in Equation 2 are related to standard deviation, σ_x and σ_y. Recall from statistics that $S_x = \sigma_x^2$. The standard deviation values are of practical importance in contaminant-plume assessment because these are related to plume size. For example, for an elliptic, symmetric contaminant plume, 99.7 percent of the mass is contained within a distance equal to 3σ from the location of the plume center of mass.

The dispersion coefficients can be estimated from a plot of spatial variance versus time by calculating the slope and dividing by two.

$$D_x = \frac{1}{2}\frac{\partial S_x}{\partial t} \tag{4a}$$

$$D_y = \frac{1}{2}\frac{\partial S_y}{\partial t} \tag{4b}$$

The primary cause of velocity variations is heterogeneity in the distribution of hydraulic conductivities. In this exercise, we use an interactive model to generate a heterogeneous aquifer and to simulate the transport of particles in the flow field. This is an advection-only model, where the velocities for each grid block are calculated with groundwater flow equations. The dispersion process is considered to be Fickian if the plot of S versus t shows a straight-line relation.

NUMERICAL MODEL PARTICLEFLOW

The model you will use for this laboratory exercise is called PARTICLEFLOW, which was developed by Dr. Paul Hsieh at the U.S. Geological Survey. The model can simulate steady-state groundwater flow and particle transport. PARTICLEFLOW is a finite-element numerical model.

The two-dimensional model PARTICLEFLOW simulates flow in a rectangular domain. A key purpose of the PARTICLEFLOW model is to illustrate how heterogeneities in hydraulic properties cause the spatial spreading of fluid particles. This spreading is analogous to macro-scale solute dispersion.

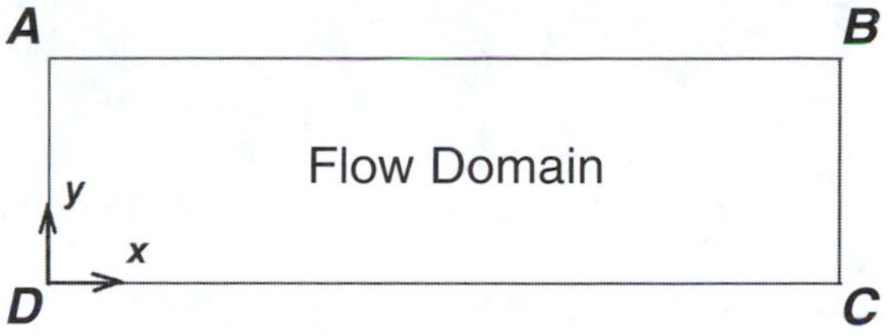

The rectangular flow domain is bounded on the left and right sides (AD and BC) by specified-head boundaries and on the top and bottom (AB and DC) by no-flow boundaries.

Assuming that the head along AD is higher then the head along BC, the average flow is from left to right. The user defines the overall dimensions of the flow field, the size of individual elements, and the hydraulic gradient. For this exercise, you will use a flow field 1000 m long by 400 m wide, an element size of 20 m, and a hydraulic gradient of 0.001.

The user can design the aquifer to be homogeneous or heterogeneous. For a homogeneous aquifer, a single value of hydraulic conductivity and porosity is assigned to the entire domain. For a heterogeneous aquifer, five different materials can be defined, each with different K and n values. These five materials can be spatially distributed in a random mix, which is the method you will use in this exercise, or they can be assigned to specific areas of the domain, which you can experiment with later on your own.

Please find the <tp1.0> folder on the CD included with this lab manual. You should copy this to an appropriate directory on your computer (for example, c:\temp), and a subdirectory will be created called <tp1.0>. In this folder, you should find a text file called <readme.txt>. Read it.

If you are working in a Windows environment, scroll down to the <pflow_win> folder (for other operating systems, refer back to the <readme> file). This contains the user's manual for the model in PDF file "ofr01-286.pdf" (USGS Open File Report 01-286; this report also includes the user's manual for the model "topodrive," which you will not use). The <pflow_win> folder also contains the PARTICLEFLOW model executable file <pflow.exe>.

Note: The model PARTICLEFLOW is a public-domain program that uses Java Runtime Environment (JRE), Version 1.1. JRE 1.1 is bundled with PARTICLEFLOW so that it can run as a stand-alone Java application. JRE is copyrighted by Sun Microsystems, Inc. Sun Microsystems has granted royalty-free distribution of JRE for the purpose of running Java applications.

PROCEDURES FOR PARTICLE-PLUME SIMULATIONS

To use the model, double-click on file "pflow.exe." The program window should open, and you are ready to start!

Task 1—Simulate Flow in a Homogeneous Aquifer

1. Click the START button to start the model simulation. In the box, use an element size of 20 m. Use 50 columns (this makes the model length = 1000 m) and 20 rows. Use an average hydraulic gradient of 0.001 (this is applied between the ends of the model domain). Click OK.

2. Click on aquifer PROPERTIES and enter the same hydraulic conductivity and porosity values for all five materials, creating a homogeneous aquifer. Use a hydraulic conductivity value of 1E-04 and a porosity value of 25.0 percent.

3. Solve for Head (click the HEAD button), using 20 contour intervals. Click COMPUTE.
   (a) ***Describe the spacing and distribution pattern of the hydraulic-head contours.***

   (b) ***How should groundwater flow lines appear?***

   Click on FLOW, and choose the "Flow path tracking" option. Click on several places along the left boundary to initiate flow lines.

4. Create an initial slug of particles by selecting the FLOW button and selecting the "Particle movement" option. Use a particle spacing of 3 m. If the program runs very slowly on your computer, increase this particle spacing.

   Next, on the model domain, draw an initial slug (i.e., a square of particles). Place the mouse cursor on the screen. Note the coordinates displayed in the lower left corner. Click at the coordinates 50,100, again at 50,300, again at 250,300, and finally, move the cursor to 250,100 and click *while pressing the CONTROL key*. Your domain should look similar to this:

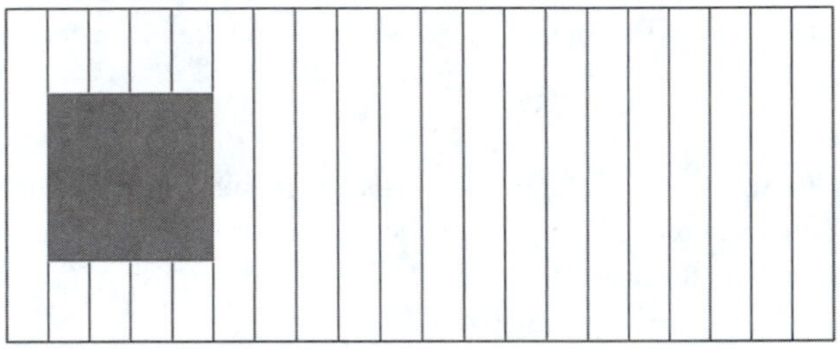

5. You will first model an aquifer with longitudinal and transverse dispersivities equal to zero; thus, dispersion will be due *only* to the aquifer heterogeneity.
   (a) ***Do you expect to see significant dispersion (spreading) of the particles?***

   Select Animation. Select the parameters as shown below. Note that the local dispersivities (both longitudinal and transverse) are set to zero.

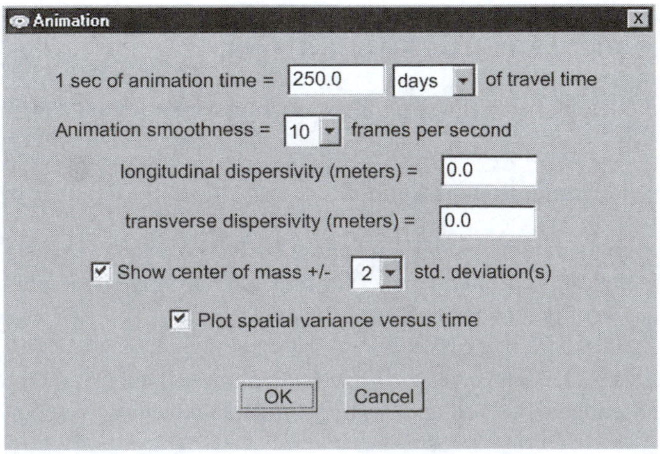

6. The model domain reappears with green and blue arrows and lines. The arrows represent the vertical and horizontal centers of mass of the particle plume, and the colored lines outside the axes represent two standard deviations (which are the square roots of the variances).

   Notice that two windows appear. One contains a plot of the horizontal spatial variance (S_{xx}) versus time, and the other vertical spatial variance (S_{YY}).

   (a) ***When do you predict the center of mass will reach the 600-m x-coordinate? Show how you reach this estimate.***

   (b) ***When do you predict the last particle will leave the end of the model at the 1000-m x-coordinate? Show how you reach this estimate.***

7. To stop or restart the simulation, simply click on the simulation screen. You may do this at any time, and at multiple times, during the simulation. If you wish to abandon the simulation, select START (or any of the other intermediate steps defined by buttons) and redefine the problem. Note that travel times can be read from the bottom of the domain simulation window.

   Use your cursor to find the location of $x = 600$ m by reading the coordinates at the bottom. Now start the simulation by clicking on the screen, being prepared to stop the simulation when $x_c = 600$ m; that is, when the x-direction center-of-mass arrow reaches 600 m. Record this time, and restart the simulation by clicking on the simulation screen.

   (a) ***Describe the movement of the plume and the spreading of the particles. Explain the spreading behavior.***

   (b) ***Report the times calculated by the model for the travel referred to in questions 6a and 6b. Were your predictions accurate? If not, explain why.***

8. Estimate the lumped-dispersion coefficients (D_x and D_y), using the information shown on the plots and using Equation 4.

Task 2—Simulate Flow in a Heterogeneous Aquifer

9. Run a second model simulation, this time using a more realistic porous medium that has random heterogeneities.

   Return to PROPERTIES, and enter variable values of *K*, as shown here. Note that *K* varies over two orders of magnitude for this simulation. Check the "Randomize" box, which will create a random distribution of *K*, using your entered values.

   (a) ***What range of soils does this simulation represent?***

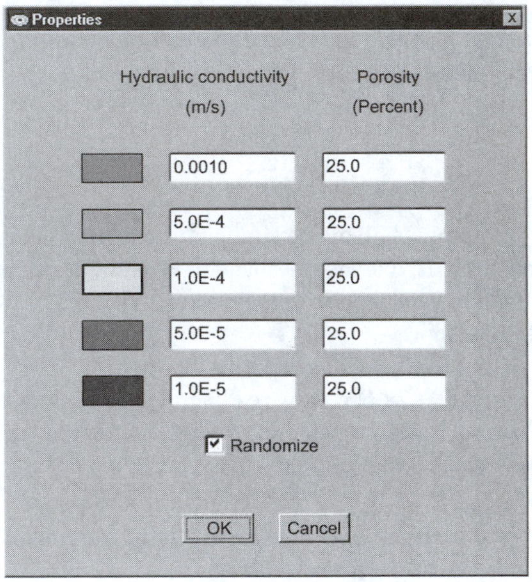

   (b) ***Describe the head distribution compared with the homogeneous case and discuss reasons for the difference.***

   (c) ***When do you predict the center of mass will reach the 600-m* x*-coordinate? Show how you reach this estimate.***

   (d) ***When do you predict the last particle will leave the model at the 1000-m* x*-coordinate? Show work.***

10. Repeat steps 3 through 8 without changing any other parameter. Be prepared to stop the simulation by clicking on the screen when the longitudinal center of mass reaches 600 m.

    (a) ***Describe the movement of the plume and the spreading of the particles. Explain the spreading behavior.***

    (b) ***Report the model calculated times for the travel referred to in questions 9c and 9d. Compare these times to those for the simulation using homogeneous* K.**

    (c) ***Is the spreading process Fickian? Explain why or why not, based on model simulation results.***

    (d) ***Estimate values for horizontal and vertical dispersion coefficients, using Equation 4.***

Task 3—Simulate Flow in a Homogeneous Aquifer, Using Dispersion Coefficients

Now consider a simulation that includes the dispersion coefficients but does not account for velocity variations due to local heterogeneity. For this task, consider the heterogeneous model from Task 2 to represent a realistic (but hypothetical) case of contaminant-plume transport in porous media. In the following task, you will compare homogeneous, lumped-dispersion-coefficient simulations to simulations of the actual spreading behavior that is due to velocity variations. Your comparisons should be based on center-of-mass travel times, travel times for the tail of the plume, horizontal and vertical spreading (e.g., the standard deviation bars in the simulation), and overall shape and behavior of the plume.

11. Estimate a single value for hydraulic conductivity based on the five K values used in the heterogeneous aquifer model.

    (a) ***What value did you select for the homogeneous* K *and why?***

    Calculate the groundwater average-linear velocity using your estimated K value.

    Estimate the dispersivities, based on your values for velocity (above) and dispersion coefficients (from 10d).

    (b) ***What are your calculated values for* α_x *and* α_y*? Show your work!***

12. Run the simulation using the values you estimated above.

    (a) ***Is the process Fickian?***

(b) ***Qualitatively describe the plume migration and the variation of the 2σ colored bars. How does the spreading compare to that for the heterogeneous simulation?***

(c) ***What does the model report for the average travel time to the 600-m x-coordinate? What does the model report for the time that the last particle exited the edge of the model at x = 1000 m?***

(d) ***What adjustments can you make to improve the ability of this homogeneous model to simulate the actual plume movement (as defined by the previous simulation in heterogeneous porous media)?***

Task 4—Improve Simulated Flow in a Homogeneous Aquifer, Using Dispersion Coefficients

13. Refine your model based on your answer to question 12d, and run another simulation.

    (a) ***Have you achieved a better approximation of the plume transport in heterogeneous media? Discuss the difficulties in using a homogeneous, lumped-parameter approach to model dispersion in heterogeneous porous media.***

14. Answer the following question.

    (a) ***Do you think that dispersion should have been included in the heterogeneous simulation as well? Why, or why not?***

15. Discuss the advantages and limitations of the following modeling approaches to simulate, understand, and predict contaminant plume movement in the real world.

    (a) Use of a lumped-dispersion coefficient and homogeneous porous media

    (b) Accounting for velocity variations and heterogeneous porous media

16. As time and interest dictate, explore the capabilities of PARTICLEFLOW. For example, one can add and investigate the influence of discrete layers or heterogeneities by mouse clicking on the square color associated with a certain hydraulic conductivity and then drawing a shape for this *K* zone in the flow domain.

LAB 16

SEYMOUR HAZARDOUS WASTE SITE I: HYDROGEOLOGIC SETTING

OVERVIEW: This series of three integrated lab exercises will provide you with a detailed knowledge of a hydrogeologic environment typical of much of the central and eastern United States. Although the hazardous nature of the site is not typical, it is, unfortunately, not rare. The exercises in these labs are typical of studies that would be conducted initially at such a site when faced with a problem of groundwater contamination.

PURPOSE: Establish the basic hydrogeology of the site by interpreting borehole data.

OBJECTIVES: Learn to work with typical geologic logs to derive subsurface information.

Make a structure contour map of the bedrock surface.

Construct two geologic cross sections to show three-dimensional variations in lithology.

Use these products to interpret and describe the hydrogeology of the site.

INTRODUCTION

Seymour Recycling Corporation operated a solvent-recovery and solvent-recycling plant at Seymour, Indiana, from 1970 to 1979. In 1979, the corporation went into bankruptcy and abandoned the location, leaving 98 large tanks and approximately 50,000 drums of organic chemicals on the site. For the next four years, an unknown amount of liquid synthetic organic compounds leaked into the soil from deteriorating drums before they were all removed in 1983. A substantial amount of these organic compounds percolated into the aquifer and threatened private water-supply wells north of the site and at least one municipal well less than a quarter of a mile to the northeast.

Site Geology

The Seymour hazardous waste site is in south central Indiana (Fig. 16.1). The general geology in this area consists of a blanket of unconsolidated periglacial deposits overlying shale bedrock. The unconsolidated sediments consist of alluvium of glacial sluiceway origin and lacustrine silts and clays. The shale bedrock is impermeable.

Hydrogeologic Maps and Sections

One of the most important methods of displaying and interpreting hydrogeologic data is through the use of maps and cross sections. These graphical representations of data also provide a way for the hydrogeologist to present information and conclusions to a nongeologist.

The information used to make cross sections and hydrogeologic maps comes chiefly from wells. When a well is drilled or a soil boring is made, a descriptive log of the strata encountered is usually made. The log is sometimes made by the well driller, in which case it is referred to as a driller's log. Because the educational level and understanding of geology varies widely from one driller to another, the quality of their well logs also varies. In most instances, a geologic log made by a geologist will provide more reliable information. Geophysical logs made using various borehole techniques may also be used to discover the nature of subsurface formations.

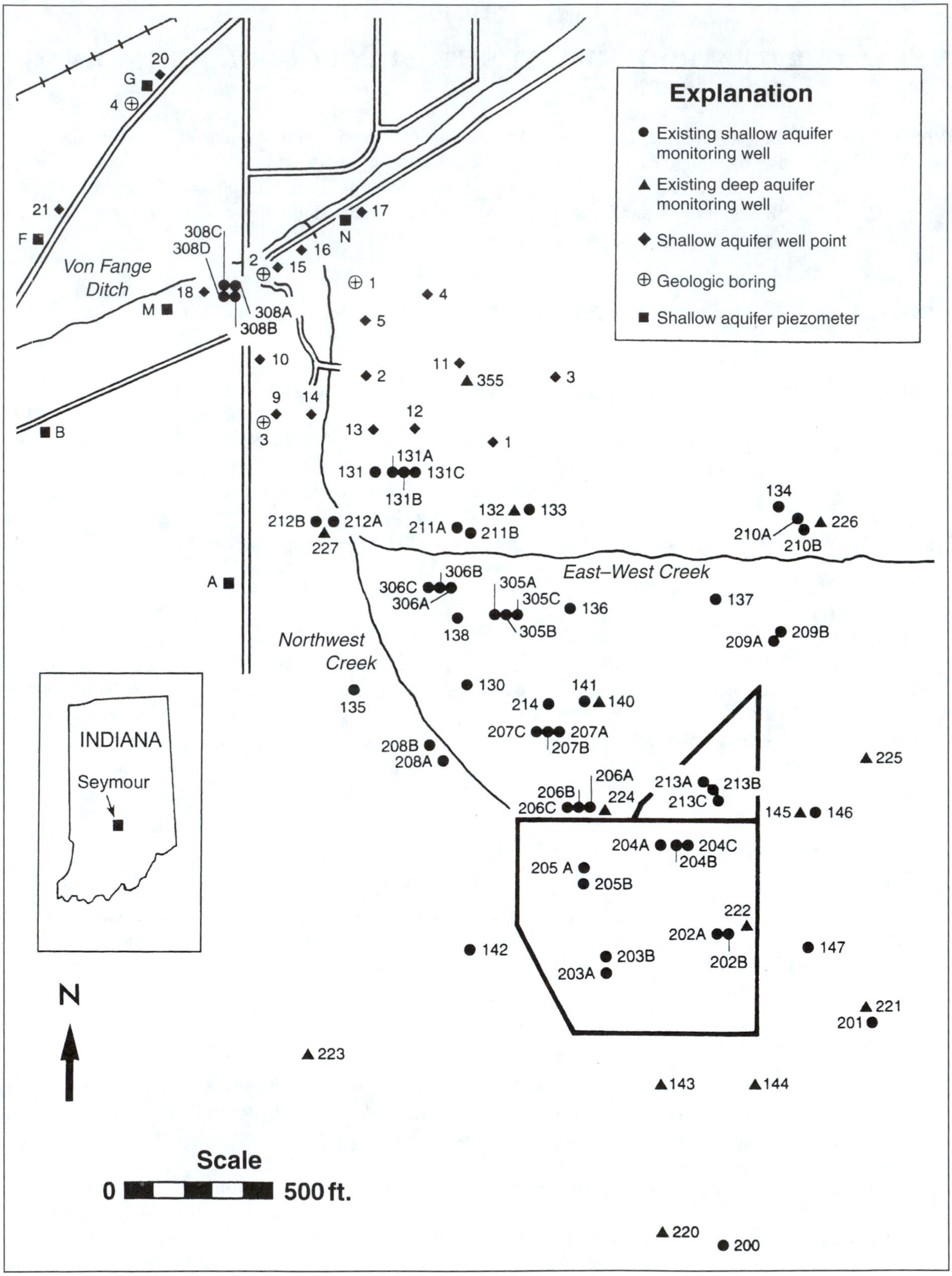

Figure 16.1—Location of Seymour Recycling Corporation site (heavy outline) at Seymour, IN (inset map).

Structure contour maps are contour maps of the surface of specific geologic units. The most common example is the bedrock contour map. The elevation of the unit at each control point is marked on a map. In some instances, the control point may indicate that the unit is known to be deeper than an indicated value, but the exact elevation is not known. Typically, the geologist will know the surface elevation of the control point, and from a well log showing the depth to the geologic unit, the elevation of the unit can be determined.

Geologic cross sections are constructed from information obtained from well logs and outcrops. Cross sections go from one control point to the next, generally following in the same direction, but not necessarily in a straight line. The distance from one control point to the next is measured on a map, and this constitutes the horizontal distance between wells on the cross section. At each control point, the well log is constructed in a vertical direction, typically with the scale in the vertical dimension much larger than the horizontal scale.

PROCEDURES

Structure Contour Map of Bedrock

1. Make a structure contour map of the bedrock surface on the site base map, Figure 16.2 on page 125.

   Data will come from the geologic logs and from some of the temporary well points. Datum for all elevations is mean sea level, and all measurements of elevation and depth are in feet.

   Use a contour interval of 2 ft.

2. Briefly describe the bedrock topography.

Geologic Cross Sections

3. Using the information in the borehole geologic logs, make two geologic cross sections.

   Cross section 1—a northeast-looking section from GB2 to GB3 to 227 to 224 to 222 to 221.

   Cross section 2—a northwest-looking section from 223 to 224 to 209 to 226.

   Show the two lines of section on Figure 16.1.

   Horizontal scale: 1 in = 500 ft; vertical scale: 1 in = 20 ft; vertical exaggeration: 25.

Hydrogeologic Interpretation

4. Using your map and sections together, consider whether or not the paleotopography was significant—that is, did it appear to influence the Quaternary sedimentation?

5. Interpreting your cross sections, describe the hydrostratigraphy of the site.

   To facilitate your interpretation, consider these questions:

   How many units are there, or does the number vary within the site?

   Is there one aquifer, or is there more than one?

   Does the geometry of the aquifer(s) change within the site?

   If there may be more than one aquifer, how do you think they might relate? Would they be clearly separated by a good confining layer, or might there be flow between them?

6. On your cross sections, outline the hydrostratigraphic units, using a different line symbol or line weight, and label the units.

GEOLOGIC LOGS OF BORINGS SEYMOUR HAZARDOUS WASTE SITE[1]

GB-1	GE (ground elevation) 562
0–6	fine silty, clayey sand (SM-SC)[1]
6–56	medium to coarse sand (SP)
56–58	silt (ML)

GB-2	GE 562
0–7	clay and silty sand (SM-CL)
7–70	fine to medium sand (SP)
70	shale bedrock

GB-3	GE 565
0–7	fine silty sand (SM)
7–60	fine to coarse sand (SP)
60–61	clay and silt (ML-CL)

GB-4	GE 566
0–7	fine to medium sand, silty (SM)
7–68	sand, medium to coarse (SP)
68–70	shale bedrock

209	GE 564.2
0–6.5	silty sand to sandy clay (SM-SC)
6.5–39.5	medium to coarse sand, clean (SP)
39.5–41	silt (ML)

220	GE 573.2
0–5.8	medium to fine sand, silty (SM-SP)
5.8–9.4	silt with fine sand (ML)
9.4–20.3	medium sand, slightly silty (SM-SP)
20.3–68	interbedded silt and clayey silt (ML-CL) with occasional fine sand layers (SM)
68–81	coarse to fine sand, some gravel (SM-SP)
81–81.5	red shale bedrock

221	GE 569.8
0–10.5	fine to medium sand, silty to clayey (SC-SP)
10.5–33	coarse to fine sand (SP-SM)
33–55.5	interbedded silt, clayey silt and sandy clay (ML-CL)
55.5–65	coarse to fine sand (SP)
65–76.6	gravel to fine sand (GW-GM)
76.6–77.1	red shale bedrock

222	GE 568.0
0–3	artificial fill
3–19.5	fine to medium sand, silty to clayey (SC-SP)
19.5–22	silt with clay (ML-CL)
22–26	fine sand (SM)
26–51.5	silt with trace of clay (ML-CL)
51.5–63.5	medium sand (SM)
63.5–73.5	medium gravel to sand (GP)
73.5–74	red shale bedrock

223	GE 570.5
0–9	sandy, silty clay (SC)
9–18.5	medium sand, slightly silty (SP)
18.5–68.5	interbedded silt and clayey silt (ML-CL)
68.5–79.5	medium to coarse sand and gravel (SP-GP)
79.5–80	shale bedrock

224	GE 564.5
0–5	fine to medium sand, silty (SM)
5–15.7	coarse to fine sand, slightly silty (SM-SP)
15.7–18	silt (ML)
18–39.5	fine to medium sand, slightly silty (SM-SP)
39.5–65.5	interbedded silt (ML) and clayey silt (ML-CL)
65.5–69.5	medium to coarse sand (SP)
69.5–70	shale bedrock

[1]Parenthetic annotations are soil designations in the commonly used Unified Soil Classification System (USCS), a system based on grain size, sorting, and plasticity (U.S. Army Engineer Waterways Experiment Station, 1960). The system is described in any basic textbook on geotechnical engineering.

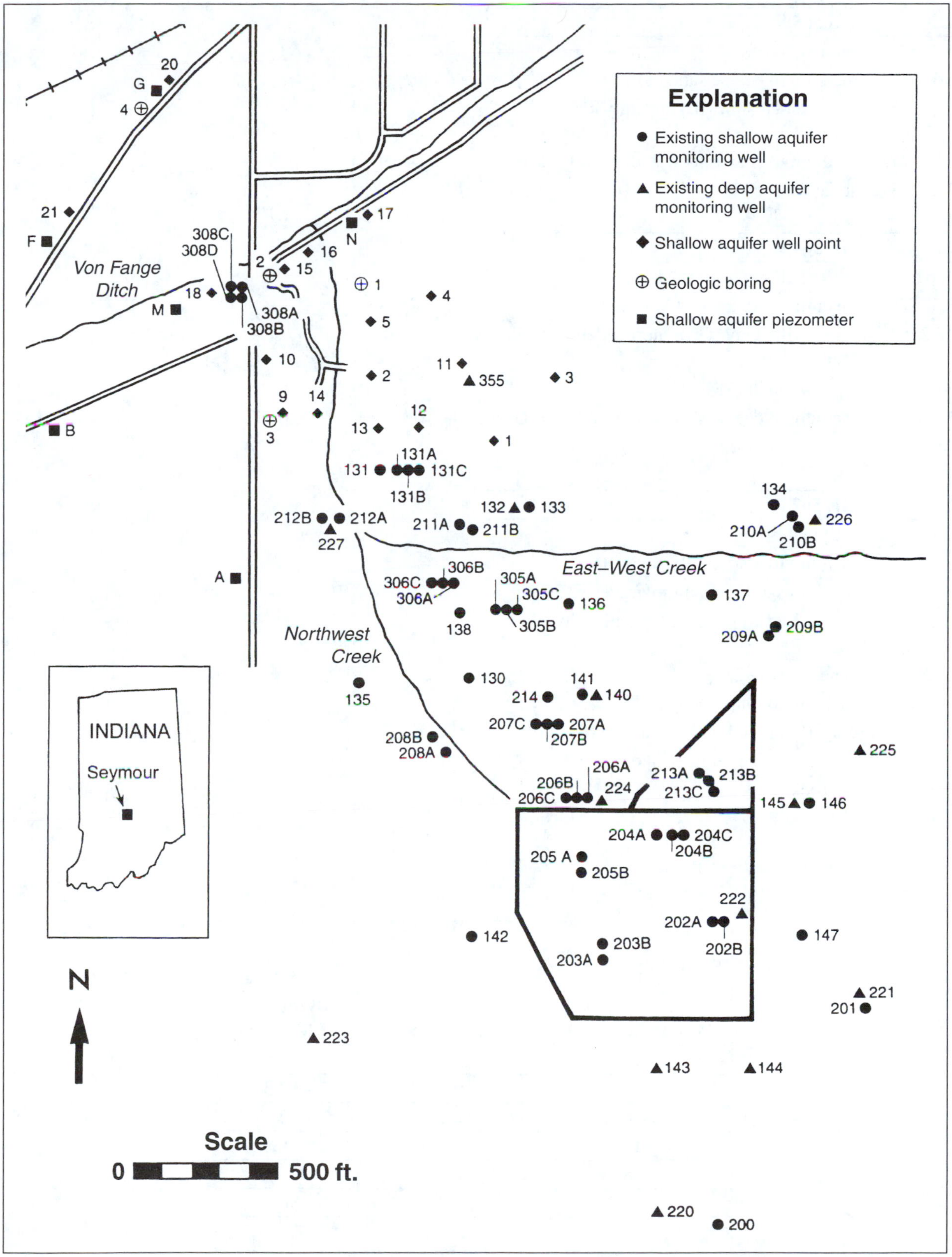

Figure 16.2—Bedrock structure contour map, Seymour hazardous waste site (contour interval 2 ft).

225	GE 565.8
0–6.5	sandy clay (SC)
6.5–43	medium to coarse sand (SP)
43–64.5	interbedded silt (ML) and silty clay (ML-CL)
64.5–77.5	medium to coarse sand with gravel (SP-GP)
77.5–78	shale bedrock

226	GE 564.9
0–6.5	silty to clayey sand (SM-SC)
6.5–37.2	medium sand, trace of silt (SP)
37.2–54	sandy, silty clay (CL) interbedded with silt (ML)
54–57	sand with silt (SM-SP)
57–74.5	medium to coarse sand with gravel (SP-GP)
74.5	shale bedrock

227	GE 562.6
0–6.5	artificial fill
6.5–9	medium to coarse silty sand (SM)
9–16.5	medium sand to medium gravel (SP-GP)
16.5–25.5	coarse to medium sand, slightly silty (SM-SP)
25.5–27	sandy clay (SC)
27–36	medium to fine sand (SP)
36–66.2	clay, plastic (CH) to clayey silt (CL-ML)
66.2–67.5	medium gravel to fine sand (GP-SP)
67.5–68	bedrock

BEDROCK SURFACE ELEVATIONS FROM TEMPORARY WELL POINTS

Well Point Number	Elevation	Bedrock Elevation
10	566.0	494.0
15	562.3	492.3
16	561.8	492.8
17	562.3	488.8
18	559.8	494
20	564	496
21	558.7	493.2

LAB 17

SEYMOUR HAZARDOUS WASTE SITE II: GROUNDWATER FLOW

OVERVIEW: This is the second in a series of three integrated lab exercises that will familiarize you with a hydrogeologic environment typical of much of the central and eastern United States. The exercises in these labs are typical of studies that would be conducted initially at such a site when faced with a problem of groundwater contamination.

PURPOSE: Determine groundwater flow at the Seymour hazardous waste site.

OBJECTIVES: Determine the number and type of aquifers at the site.

Establish the nature and extent of a hydrostratigraphic unit by making an isopach map.

Map the potential distribution in the aquifer(s), and determine groundwater flow directions.

HYDROGEOLOGIC MAPS

The water level measured in a well is a measure of the hydraulic head (or potential) in the aquifer at the screened zone of the well. Wells that set screens into more than one hydrostratigraphic unit have a water level that reflects the combination of the water levels of the different screened zones; the data from such wells are difficult to use. Wells in unconfined aquifers show the position of the water table. Wells screened into confined aquifers indicate the position of the potentiometric surface. Clusters of several wells at the same location, but screened at different depths, can show the vertical component of groundwater movement.

Following are two additional types of hydrogeologic maps that are made.

Isopach maps show the thickness of a particular geologic unit. They are constructed by finding the thickness of the unit at each of the control points, either outcrops or well logs. The thickness at each control point is located on a map, and a contour map is constructed on the basis of the hydrostratigraphic unit thickness.

Potentiometric surface maps show the distribution of head, or potential, throughout an aquifer. They are made by plotting the locations of wells on a base map, and water levels, measured within a short period of time, are plotted at the well locations. For confined aquifers, the potentiometric surface is contoured from water levels in wells screened in the same aquifer, and contour lines may be drawn without regard to the influence of surface hydrologic features.

For unconfined aquifers, the potentiometric surface is the water table. Water levels are measured within a short period of time (within a few days, with no major precipitation during the measurement period). Streams, lakes, and rivers that might be linked to the water table are also mapped, and contours must be drawn keeping in mind how these surface water features may impact the water table.

PROCEDURES

Define Aquifers

1. In the last lab, you made an interpretation of the hydrostratigraphy of the site. Did you suggest there is one or two aquifers? If more than one, were the aquifers part of the same flow system—that is, are they hydraulically connected—or are there two independent flow systems?

2. If you could measure standing water levels in two wells side by side, one open at depth and the other only shallow, how would this information help you answer this question?

   (a) If the two levels were exactly the same, what would you infer?

   (b) If the level in the deep well were higher, what would you infer?

3. Return to the cross sections you made in the last lab, and plot on them the water levels recorded April 8, 1990, in each of the deep wells (Water Levels in Deep Wells, April 8, 1990 [p. 132]). Use a red line to connect the water levels. Note that for GB-2 at the northwest end of the section, you can use data from 308-D nearby. Similarly, you can get information from shallow wells that were drilled and monitored next to each of the deep wells. Referring to the site map, you can see at the northwest end, for example, 308-A, a shallow well, is adjacent to 308-D.

   Plot data from Water Levels in Shallow Wells, April 8, 1990 (p. 134), on the cross sections also, using blue for these measurements.

   Considering the relationship of the upper potentiometric surface to the stratigraphy, is the upper aquifer confined, unconfined, or semiconfined?

4. Describe and interpret the relationship of the two potentiometric lines.

Construct an Isopach Map of the Confining Layer

1. Once you have established the relationship of the upper and lower aquifers and your cross section 1 shows some limitation to the separateness of these aquifers, it would be instructive to determine the three-dimensional nature of the confining layer. In addition to the geologic logs, some information is available from a temporary well point program (p. 132). Using both of these data sources, construct an isopach map for the confining layer on the site base map, Figure 17.1.

2. Describe the geometry of the confining layer, and suggest an explanation for it.

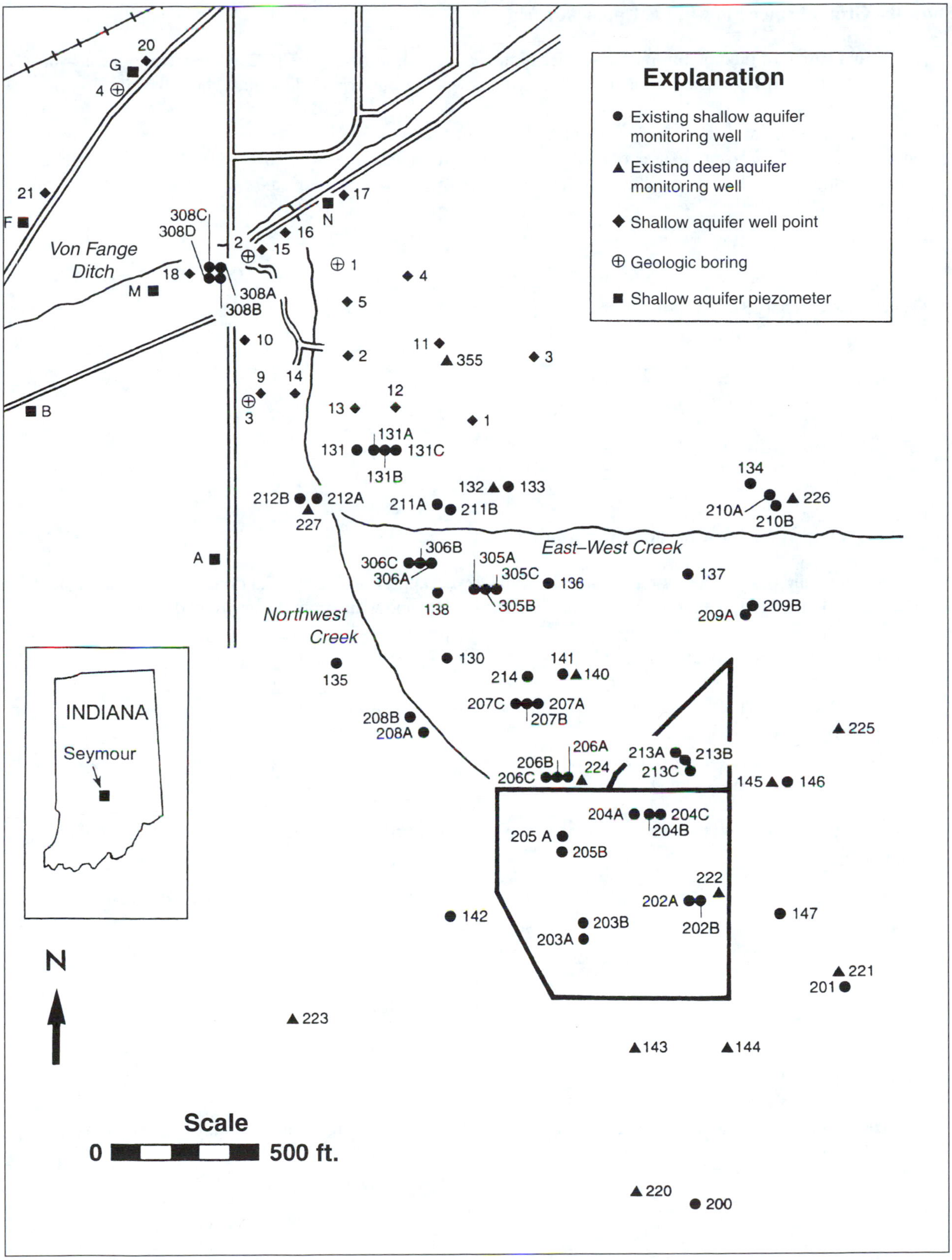

Figure 17.1—Isopach map of the confining layer, Seymour hazardous waste site (contour interval, 5 ft).

Show the Groundwater Flow in the Upper Aquifer

1. Construct a map of the potentiometric surface of the upper aquifer on Figure 17.2, using a contour interval of 1 ft. With blue arrows, draw sufficient flow lines to show the groundwater flow throughout the site.

2. One could use this map to estimate actual travel paths of groundwater. For example, where would you expect groundwater to flow from the area of well 204? Draw a green flow line from 204B to the Von Fange Ditch. Where along the ditch (nearest which well) would you expect this flow to occur?

3. Where does recharge to this aquifer probably occur?

Show the Groundwater Flow in the Lower Aquifer

1. Construct a map of the potentiometric surface of the lower aquifer on Figure 17.3 on page 133, using a contour interval of 1 ft. With red arrows, draw sufficient flow lines to show the groundwater flow throughout the site.

2. Where does recharge probably occur?

3. Compare the heads from the two potentiometric maps north of the Von Fange Ditch. What do you think is happening in this area?

Summary of Hydrogeology

Summarize, in a few paragraphs, the hydrogeology of the site, making reference to your maps and sections.

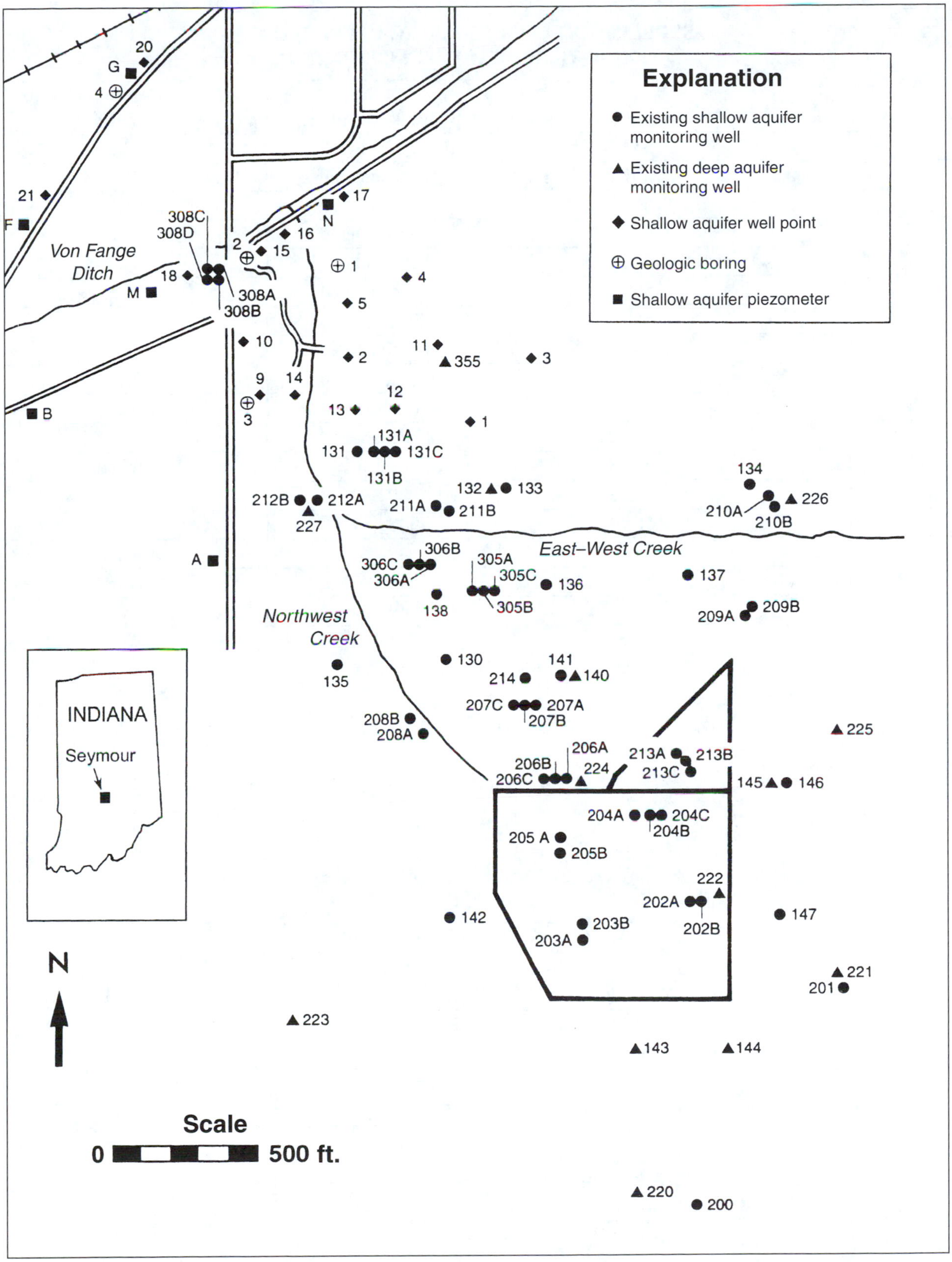

Figure 17.2—Potentiometric surface and groundwater flow in the upper aquifer (contour interval, 1 ft).

SUPPLEMENTARY CONFINING LAYER DATA FROM TEMPORARY WELL POINTS			
Well Point	Ground Elevation	Bedrock Elevation	Top of Confining Layer Elevation
1	552	not encountered	502
2	561.7	not encountered	501.7
3	563.1	not encountered	503.1
4	562.7	not encountered	503.7
5	560.9	not encountered	503.9
9	565.5	not encountered	505.5
10	566.0	494.0	506.0
11	562.7	not encountered	501.7
12	562.2	not encountered	503.2
13	562.4	not encountered	507.4
14	562.2	not encountered	507.2
15	562.3	492.3	not found
16	561.8	492.8	not found
17	562.3	488.8	not found
18	559.8	494	not found
20	564	496	not found
21	558.7	493.2	not found

WATER LEVELS IN DEEP WELLS April 8, 1990	
Well	Water-Level Elevation
132	556.80
140	556.65
145	556.49
220	551.59
221	550.69
222	551.15
223	550.82
224	556.32
225	556.64
226	556.91
227	555.34
308D	555.43
355	556.62

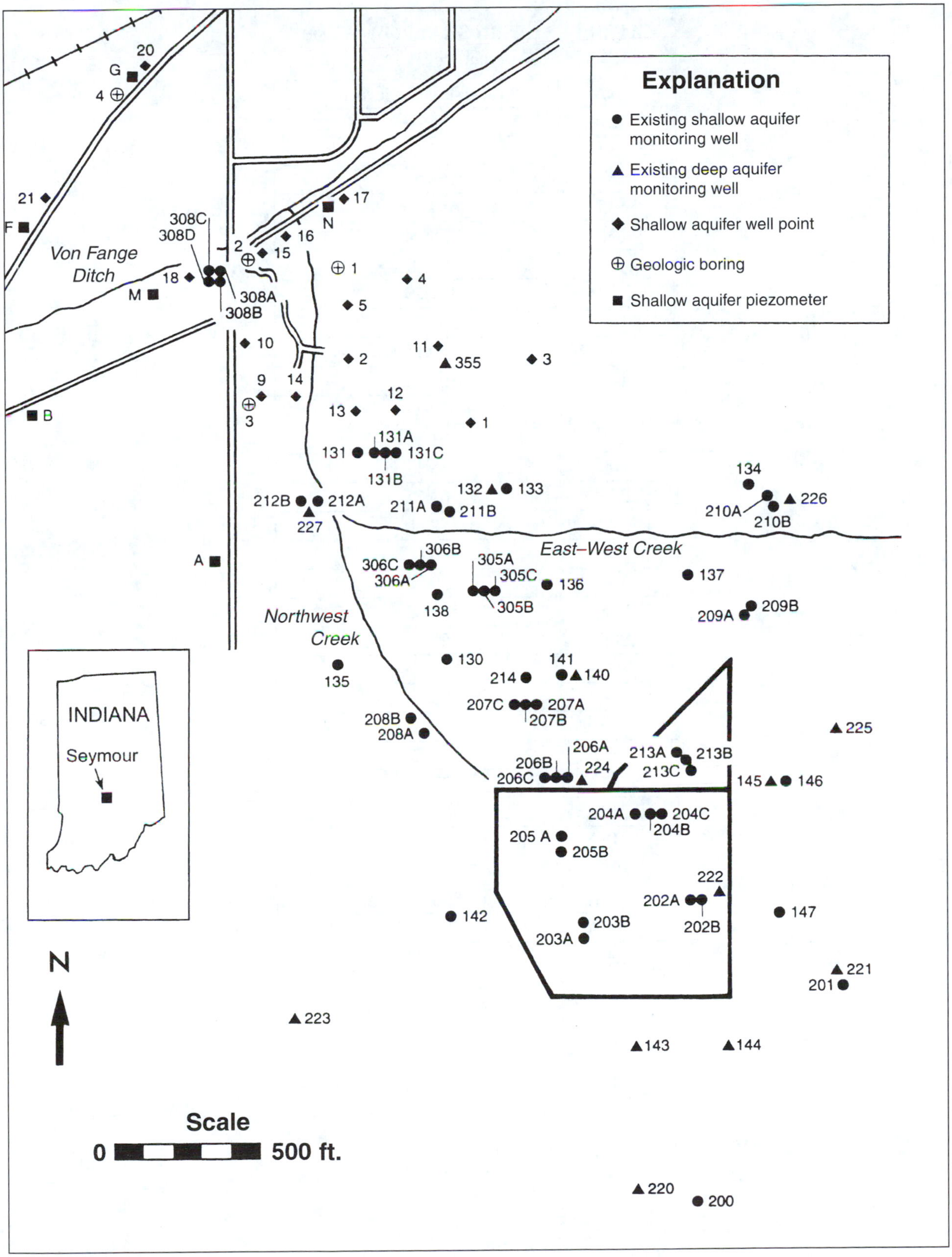

Figure 17.3—Potentiometric surface and groundwater flow in the lower aquifer (contour interval, 1 ft).

WATER LEVELS IN SHALLOW WELLS April 8, 1990	
Well	**Water-Level Elevation**
131	556.40
133	557.37
134	558.55
135	557.50
136	558.04
137	559.01
138	557.79
141	558.79
142	560.18
144	563.67
146	560.31
147	561.15
200	570.11
201	561.75
202A	559.80
203A	560.80
204A	559.78
205A	559.82
206A	559.16
207A	558.46
208A	558.43
209A	559.08
210A	558.72
211A	557.04
212A	556.78
213A	559.41
305A	557.73
306A	556.54
308A	555.43
A	557.92
B	554.83
F	554.22
G	554.77
M	555.03
N	557.10

LAB 18

SEYMOUR HAZARDOUS WASTE SITE III: GROUNDWATER CONTAMINATION

OVERVIEW: This is the third in the series of three integrated lab exercises that will familiarize you with a hydrogeologic environment typical of much of the central and eastern United States. The exercises in these labs are typical of studies that would be conducted initially at such a site when faced with a problem of groundwater contamination.

PURPOSE: Determine groundwater contamination at the Seymour hazardous waste site.

OBJECTIVES: Map the extent of groundwater contamination at the site.

Compare actual path of contaminant flow with your earlier estimate.

Estimate the nature and extent of future aquifer contamination.

Explore possible mitigation and remediation procedures.

Groundwater Contamination

Some of the wells in and around the Seymour Recycling Corporation site were sampled within a seven-month period in 1984–1985. Most of these wells were resampled again in 1989–1990, along with a series of well points that were put in to take additional samples. The samples were analyzed for organic compounds, using gas chromatography and mass spectroscopy. Analytical results for one of these compounds, tetrahydrofuran, are tabulated on the next page.

1. Using data from the earlier survey, construct a map showing the plume of contamination due to tetrahydrofuran on Figure 18.1, page 137. Use equal-concentration lines (contour lines) of 10, 50, 100, 500, 1000, 5000, 10,000, and 50,000 μg/L.

   Describe the distribution of the tetrahydrofuran. Is it what you would have anticipated?

   Based on the results from this first survey, where would you like additional information?

TETRAHYDROFURAN IN GROUNDWATER AT Seymour Hazardous Waste Site		
	Tetrahydrofuran, μg/L	
Well	1984–1985	1989–1990
131C	—	3100
133	—	ND-50[1]
136	—	290
141	—	2200
143	—	ND-5
145	—	ND-5
200	ND	—
201	ND	—
202A	30,000	6000
203A	9000	2500
204B	20,000	37,000
205A	10,000	ND-1000
206A	12,000	6200
207A	52,000	5900
208	30	—
209A	ND	ND-10
210	ND	—
211A	530	5700
212A	ND	ND-10
213A	7000	ND-10
214	—	2100
305B		2200
306C		9800
WP-1		2000
WP-2		2
WP-3		ND
WP-4		2
WP-8		ND
WP-9		ND
WP-10		ND
WP-11		ND
WP-12		15
WP-13		12
WP-14		6
WP-15		5
WP-16		ND
WP-17		ND
WP-18		17
WP-20		ND
WP-21		ND

Note: [1]ND = none detected, unknown detection limit; ND-50 = none detected, 50 μg/L detection limit; — = no report; WP = well point.

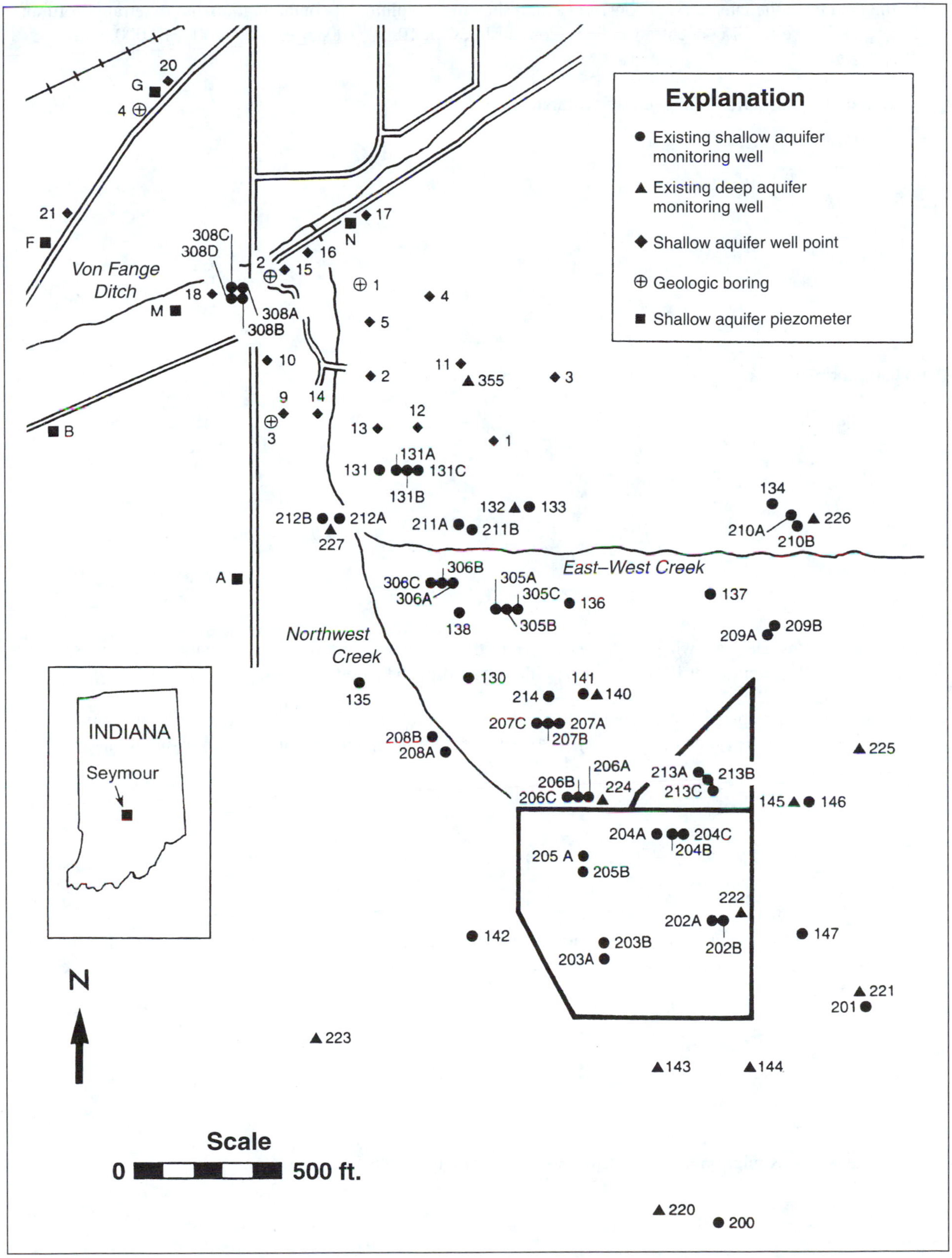

Figure 18.1—Map of tetrahydrofuran concentration (μg/L), Seymour hazardous waste site, 1984–1985.

2. Using data from the later survey, construct a map showing the plume of contamination due to tetrahydrofuran on Figure 18.2. Use equal-concentration lines (contour lines) of 10, 50, 100, 500, 1000, 5000, 10,000, and 50,000 μg/L.

   Describe the distribution of the tetrahydrofuran.

   What information can you interpret from these two maps combined?

3. Using your maps, estimate where the main source of the tetrahydrofuran contamination occurred, and show the general path of groundwater flow from this point as traced by the tetrahydrofuran.

4. How does this path compare with your earlier estimated path of groundwater flow? Refer back to your work in Lab 17 under *Groundwater Flow in Upper Aquifer*, and to the green flow line you drew on the map in Figure 17.2.

5. We don't know for sure if the tetrahydrofuran contamination occurred at a single point or over some finite area. Assuming that it was a point source at or near well 204, would you expect the contaminant to be restricted to a single flow line?

6. What mechanisms might operate to disperse the contaminants from a line into a plume?

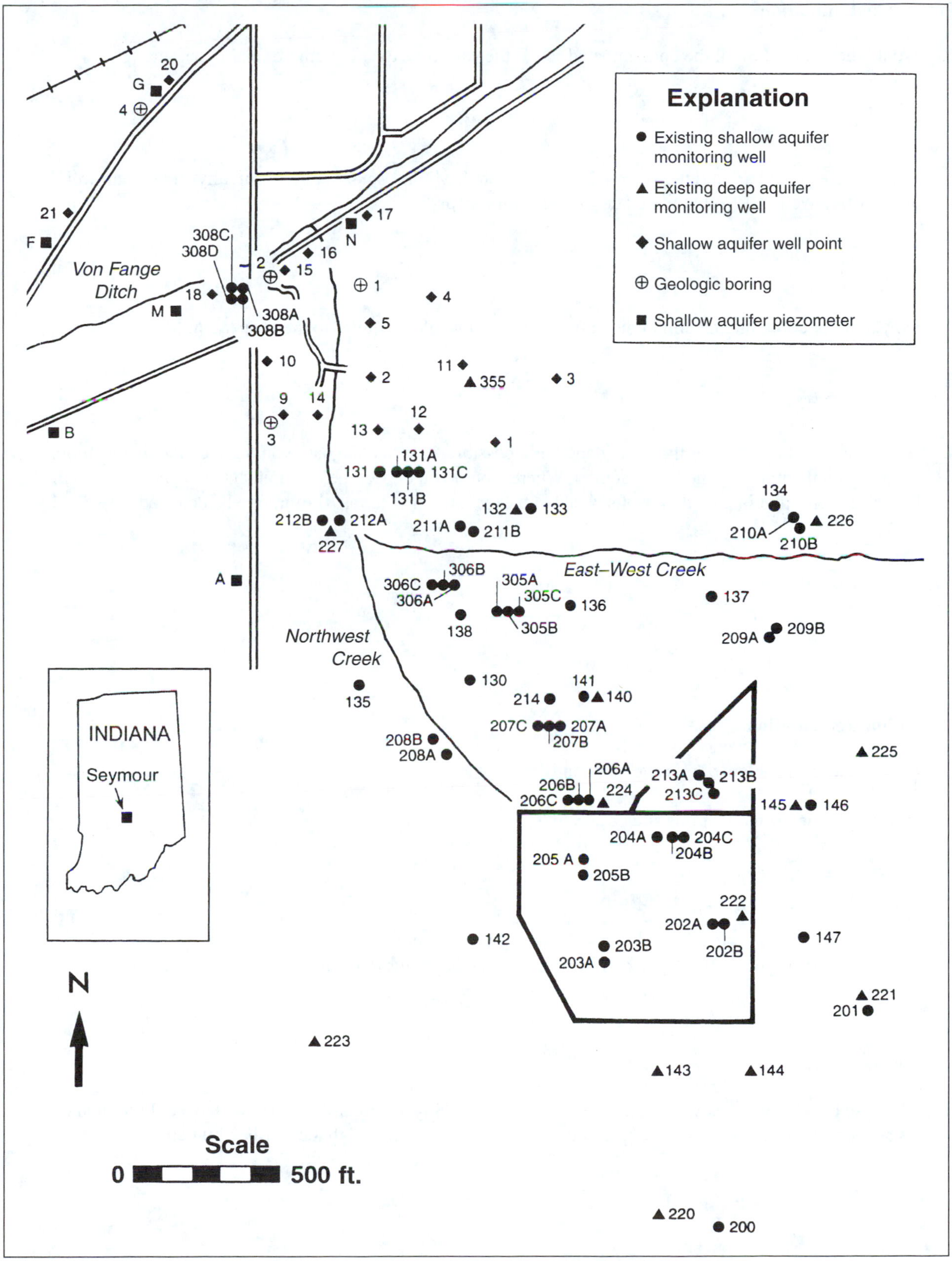

Figure 18.2—Map of tetrahydrofuran concentration (μg/L), Seymour hazardous waste site, 1989–1990.

Future Contamination

1. Would the removal of all the contaminated soil at the site solve the problem? Explain your answer.

2. All the wells in and around the site are being monitored. Might contaminants eventually show up in well 209A? Well 212A? Well 201? What would determine if this occurs?

3. Do you think contaminants might move from the upper shallow aquifer to the deeper aquifer?

4. Assuming contamination of the deeper aquifer is possible, describe the path you think contaminants might follow from the upper to the lower aquifer. Where would you expect the contaminants to appear first in the lower aquifer—that is, in which wells would you expect the first tetrahydrofuran to be detected?

Mitigation/Remediation

1. How might the problem at this site be diminished?

2. How might the problem at this site be solved?

Summary

Write a brief summary of groundwater contamination at the Seymour Recycling Corporation site.

Review

Read the case history of the Seymour site (Fetter, 2001, pp. 428–436). Realizing that you have only limited data, discuss any significant differences between your evaluation of the site and that described by Fetter.

LABS 19–21

GROUNDWATER BASIN ANALYSIS

PURPOSE: Learn to integrate geologic and hydrologic data to interpret the groundwater conditions in a basin.

Note: This is a single lab that will extend over a three-week period.

Anderman-Lee and Associates
Water Resources Engineers

Dear Hydrogeologist:

We would like to retain your services as a consultant in a groundwater study for the Lost Valley Irrigation District. They have retained us to do an evaluation of a groundwater basin in southern California.

The technical portion of the study is described in the attached document, *Groundwater Analysis of the Lost Valley Basin*. We are asking you to make an interpretation of available hydrographic data and to prepare a final technical report based on this document.

In addition, we would like a summary report that can be distributed to individual members of the district. This report should be three double-spaced pages (or less), including a sketch map and any relevant tables. The focus of the report should be the future of irrigation in the basin and the problems to be expected in future development. Recognize that this report is to be read by farmers and ranchers who are knowledgeable about water and irrigation, but who cannot be expected to understand scientific or engineering terms.

We look forward to receiving your final technical report and summary report.

Sincerely,

Evan Anderman

GROUNDWATER ANALYSIS OF THE LOST VALLEY BASIN

OBJECTIVES: Determine the aquifer conditions in the Lost Valley basin.

Determine recharge, discharge, and the direction of groundwater flow.

Determine the relationship between groundwater and surface water.

Evaluate the future of the groundwater basin.

PROCEDURES

1. Read through the material available, especially the section on *Results and Discussion*, before starting.
2. Map the potentiometric surface(s) using a 10-foot contour interval: If potentiometric surface is a simple water table, use blue pencil; if the potentiometric surface is related to a confined aquifer, use red pencil.
3. Mark areas of recharge and discharge, and show with colored arrows the direction of groundwater flow.
4. Include an explanation with the map.
5. Use the test hole information to construct a generalized cross section.
6. Analyze the temporal and spatial variations of stream flow.

DESCRIPTION OF AREA

The map in Plate 1 on page 151 represents a portion of a basin that is filled with unconsolidated sediments. Granitic bedrock crops out in the southwestern and northeastern parts of the map areas (above the 1300-foot contours). The region is arid, with an average annual precipitation of less than 6 inches. The only perennial stream is shown as a solid black line. The dash-dot lines represent dry desert washes, although the lower reaches of these usually contain water. Water levels in wells represent non-pumping conditions.

Irrigation by groundwater occurs in the following areas: sections 24 through 36, T5S, R4E; all sections of T5S, R5E; sections 7, 8, 9, and 16 through 36 in T5S, R6E; all sections in T6S, R4E; all sections in T6S, R5E; and all sections in T6S, R6E. Irrigated fields and meadows are more or less uniformly distributed through this area. About 1 percent of the total area subject to irrigation is actually under irrigation during any one year.

All irrigation water comes from gravel-packed, large-diameter irrigation wells, which are pumped continuously through the 6-month growing season and shut down for the other 6 months. All other wells in the area have nonperforated, open-end casing and use negligible amounts of groundwater. Neglect precipitation onto, and evaporation from, the map area. Do not neglect transpiration of irrigated lands.

STREAM DISCHARGE

Average monthly stream discharge, in cfs:

Station	Oct	Nov	Dec	Jan	Feb	Mar	Apr	May	Jun	Jul	Aug	Sep
Gage 1	0	3	15	20	15	10	10	5	2	1	0	0
Gage 2	0	0	5	15	8	2	1	0	0	0	0	0
Gage 3	6	5	8	17	10	3	6	7	6	4	3	8

TEST HOLES

DRILLER'S LOGS

TH 1

0–10	topsoil
10–150	gravel and sand
150–400	gravel
400–850	sand and fine gravel
850–900	sand
900–1200	gravel and sand

TH 2

0–5	topsoil
5–280	fine sand
280–295	clay
295–390	sand
390–560	clay and fine sand
560–630	blue clay
630–950	sand and fine gravel

TH 3

0–5	topsoil
5–25	fine sand
25–150	sandy clay
150–600	clay
600–790	blue clay
790–1000	fine sand

TH 4

0–5	topsoil
5–40	sand
40–125	clay
125–130	sand
130–300	clay
300–650	blue clay
650–900	fine sand
900–1000	clay and sand

TH 5

0–10	topsoil
10–50	sand and gravel
50–90	clay
90–120	fine sand
120–690	clay
690–800	fine sand

RESULTS, DISCUSSION, AND SUMMARY

1. Describe the aquifer(s)—number, distribution, thickness, lithology.
2. Where does groundwater flow out of the map area? Support your answer.
3. Describe influent/effluent reaches of the river.
4. Is groundwater tributary or nontributary? Support your answer.
5. Describe the temporal nature of recharge to the aquifer(s).
6. Describe the temporal nature of discharge from the aquifer(s).
7. Does the present irrigated acreage appear to suffer from a shortage of water? Support your answer.
8. Describe future aquifer conditions in the basin if irrigation practices continue unchanged.
9. Write an executive summary report (described in cover letter).

REFERENCES

Bedient, P.B., Rifai, H.S., and Newell, C.J., 1999, Ground Water Contamination, Transport, and Remediation: Upper Saddle River, NJ, Prentice Hall.

Bouwer, H., 1989, The Bouwer and Rice Slug Test—an update: *Ground Water*, Vol. 27, No. 3, pp. 304–309.

Bouwer, H., and Rice, R.C., 1976, A Slug Test for Determining Hydraulic Conductivity of Unconfined Aquifers with Completely or Partially Penetrating Wells: *Water Resources Research*, Vol. 12, No. 3, pp. 423–428.

Colorado Land Use Commission, 1974, Snow Depth, Colorado, 1974: Denver, Colorado Land Use Commission, scale 1 : 500,000.

Dawson, K.J., and Istok, J.D., 1991, Aquifer Testing: Chelsea, MI, Lewis Publishers.

de Josselin de Jong, G., 1958, Longitudinal and Transverse Diffusion in Granular Deposits: *Transactions American Geophysical Union*, Vol. 39, No. 1, p. 67.

Domenico P.A., and Schwartz, F.W., 1998, *Physical and Chemical Hydrogeology*, 2nd edition: New York, John Wiley and Sons.

Emery, P.A., Boettcher, A.J., Snipes, R.J., and McIntyre, H.J., Jr., 1971, Hydrology of the San Luis Valley, South-central Colorado: *U.S. Geological Survey Hydrologic Investigations Atlas HA-831*, scale 1 : 250,000, text.

Fetter, C.W., 1994, Applied Hydrogeology, 3rd edition: New York, Macmillan.

Fetter, C.W., 1999, Contaminant Hydrogeology, 2nd edition: Upper Saddle River, NJ, Prentice Hall.

Fetter, C.W., 2001, *Applied Hydrogeology*, 4th edition: Upper Saddle River, NJ, Prentice Hall.

Freeze, R.A., and Cherry, J.A., 1979, *Groundwater*: Englewood Cliffs, NJ, Prentice-Hall.

Freyberg, D.L., 1986, A Natural Gradient Experiment on Solute Transport in a Sand Aquifer 2—Spatial Moments and the Advection and Dispersion of Non-reactive Tracers: *Water Resources Research*, Vol. 22, pp. 2031–2046.

Grubb, H.F., 1998, Summary of Hydrology of the Regional Aquifer Systems, Gulf Coastal Plain, South–Central United States: *U.S. Geological Survey Professional Paper 1416-A*, 61p.

Hem, J.D., 1959, Study and Interpretation of the Chemical Characteristics of Natural Water: *U.S. Geological Survey Water-Supply Paper 1473*.

Hem, J.D., 1970, Study and Interpretation of the Chemical Characteristics of Natural Water: *U.S. Geological Survey Water-Supply Paper 1473*, 2nd ed.

Hillier, D.E., and Schneider, P.A., Jr., 1979, Well Yields and Chemical Quality of Water from Water-Table Aquifers in the Boulder–Fort Collins–Greeley Area, Front Range Urban Corridor, Colorado: *U.S. Geological Survey Miscellaneous Investigations Map I-855-J*, scale 1 : 100,000.

Hillier, D.E., Schneider, P.A., Jr., and Hutchinson, E.C., 1983, Well Yields and Chemical Quality of Water from Water-Table Aquifers in the Greater Denver Area, Front Range Urban Corridor, Colorado: *U.S. Geological Survey Miscellaneous Investigations Map I-856-J*, scale 1 : 100,000.

Hsieh, P.A., 2001, TOPODRIVE and PARTICLEFLOW—Two Computer Models for Simulation and Visualization of Groundwater Flow and Transport of Fluid Particles in Two Dimensions: *U.S. Geological Survey Open-File Report 01-286*.

Hvorslev, M.J., 1951, Time Lag and Soil Permeability in Groundwater Observations: *U.S. Army Corps of Engineers Waterways Experiment Station, Bulletin 36*.

Jacob, C.E., 1950, Flow of Groundwater, in Rouse, H., ed., *Engineering Hydraulics*: New York, John Wiley, pp. 321–386.

Kohler, M.A., Nordenson, T.J., and Baker, D.R., 1959, Evaporation Maps for the United States: *U.S. Weather Bureau Technical Paper No. 37*, 13 pp. [5 plates].

League of Women Voters of Colorado, 1975, Colorado Water: Denver, League of Women Voters of Colorado.

Lee, Keenan, 1969, Infrared Exploration for Shoreline Springs: A contribution to the Hydrogeology of Mono Basin, California: Stanford University, unpublished Ph.D. dissertation.

Lohman, S.W., 1979, Ground-Water Hydraulics: *U.S. Geological Survey Professional Paper 708*, 70 pp., 9 plates.

McConaghy, J.A., Chase, G.H., Boettcher, A.J., and Major, T.J., 1964, Hydrogeologic Data of the Denver Basin, Colorado: *Colorado Water Conservation Board Basic-Data Report No. 15*.

Meyer, R.R., and Turcan, A.N., Jr., 1955, Geology and Ground-Water Resources of the Baton Rouge Area, Louisiana: *U.S. Geological Survey Water-Supply Paper 1296*.

Murray, C.R., and Reeves, E.B., 1972, Estimated Use of Water in the United States in 1970: *U.S. Geological Survey Circular 676*.

Neuman, S.P., 1975, Analysis of Pumping Test Data from Anisotropic Unconfined Aquifers Considering Delayed Gravity Response: *Water Resources Research*, Vol. 11, No. 2, pp. 329–342.

Renken, R.A., 1998, Ground Water Atlas of the United States, Segment 5, Arkansas, Louisiana, Mississippi: *U.S. Geological Survey Hydrologic Investigations Atlas 730-F*.

Robinson, T.W., 1958, Phreatophytes: *U.S. Geological Survey Water-Supply Paper 1423*.

Robson, S.G., 1987, Bedrock Aquifers of the Denver Basin, Colorado—A Quantitative Water-Resources Appraisal: *U.S. Geological Survey Professional Paper 1257*.

Robson, S.G., 1989, Alluvial and Bedrock Aquifers of the Denver Basin—Eastern Colorado's Dual Groundwater Resource: *U.S. Geological Survey Water-Supply Paper 2302*.

Robson, S.G., and Banta, E.R., 1995, Ground Water Atlas of the United States, Segment 2, Arizona, Colorado, New Mexico, Utah: *U.S. Geological Survey Hydrologic Investigations Atlas 730-C*.

Romero, J.C., 1976, Ground Water Resources of the Bedrock Aquifers of the Denver Basin, Colorado: *Denver, Colorado Department Natural Resources, Division of Water Resources*.

Soil Conservation Service, 1972, Natural Vegetation, Colorado, October 1972: *Soil Conservation Service Map M7-E-22390*, scale 1 : 1,500,000.

Sun, R.J., and Johnson, R.H., 1994, Regional Aquifer-System Analysis Program of the U.S. Geological Survey, 1978–1992: *U.S. Geological Survey Circular 1099*.

Theis, C.V., 1935, The Relation Between the Lowering of the Piezometric Surface and the Rate and Duration of Discharge of a Well Using Groundwater Storage: *Transactions American Geophysical Union*, Vol. 16, pp. 519–524.

Tweto, Ogden, comp., 1979, Geologic Map of Colorado: *U.S. Geological Survey in Cooperation with Colorado Geological Survey*, scale 1 : 500,000.

U.S. Army Engineer Waterways Experiment Station, 1960, The Unified Soil Classification System: *Technical Memorandum No. 3-357*.

Weiss, J.S., 1992, Geohydrologic Units of the Coastal Lowlands Aquifer System, South-Central United States: *U.S. Geological Survey Professional Paper 1416-C*.

Winograd, I.J., and Thordarson, William, 1975, Hydrogeologic and Hydrochemical Framework, South-Central Great Basin, Nevada-California, with Special Reference to the Nevada Test Site: *U.S. Geological Survey Professional Paper 712-C*.

Appendix 1

HYDROGEOLOGY SYMBOLS

A area
α compressibility of aquifer; dispersivity
β compressibility of water
b thickness of aquifer
C shape constant; concentration
d grain diameter
D dispersion coefficient
γ specific weight of water
θ incidence angle (σ of Fetter)
f_{OC} mass fraction of organic carbon
g gravitational acceleration
h head
h_p pressure head
k intrinsic permeability (K_i of Fetter)
K hydraulic conductivity
K_x hydraulic conductivity parallel to bedding (K_h of Fetter)
K_z hydraulic conductivity perpendicular to bedding (K_v of Fetter)
K_D soil–water partition coefficient
K_{OC} organic carbon partitioning coefficient
1 length
λ first-order degradation rate coefficient
M moment (spatial or temporal, in context)
μ dynamic viscosity (also, as prefix *micro*-, 10^{-6})
n porosity
N_R Reynolds number (R of Fetter)
n_d number of potential drops (f of Fetter)
n_s number of stream tubes (p of Fetter)
P pressure
Q discharge
r radius
r_i radius of influence of well
R retardation factor
ρ density
ρ_b bulk density
ρ_s grain (solid) density
s drawdown ($h_0 - h$ of Fetter)
S storativity or storage coefficient; variance
S_r specific retention
S_s specific storage
S_y specific yield
σ_e effective stress
t time
T transmissivity
v velocity of groundwater (V_x, *average linear velocity* of Fetter)
v_c velocity of contaminant
V specific discharge (or Darcy velocity)(v of Fetter)
V_b bulk volume ($=V_t$)
V_t total volume ($=V_b$)
V_v volume of voids
V_s volume of solids
V_w volume of water
w width of aquifer
φ potential (*force potential* of Fetter)
x distance parallel to bedding
y distance perpendicular to bedding
z elevation head

Appendix 2

SOME RULES OF THUMB FOR HYDROGEOLOGY

Evaporation	$\frac{E_{v_{\text{May-Oct}}}}{E_{v_{\text{annual}}}} \approx 0.7$	$\frac{E_{v_{\text{lake}}}}{E_{v_{\text{pan}}}} \approx 0.7$

Precipitation: 10 in. snow $\approx$ 1 in. rain

Runoff: $vel_{\text{ave}} \approx vel_{0.6\text{d}} \approx 0.85 vel_{\text{surface}} \approx \frac{vel_{0.2\text{d}} + vel_{0.8\text{d}}}{2}$

Water chemistry:

$$\text{TDS}_{[\text{epm}]} \approx \frac{\text{specific conductance } (\mu\text{mhos})}{100}$$

$$\text{TDS}_{(\text{mg/l})} \approx 0.7 \times \text{specific conductance } (\mu\text{mhos})$$

Water quantity

- $1\ \text{ft}^3 = 7.5$ gal
- $1\ \text{af} = 43{,}560\ \text{ft}^3$
- $1\ \text{cfs} = 2\ \text{af/d} = 450$ gpm

Household supply:	60 gallons per day per capita (gpdpc)
Domestic supply:	100 gpdpc
Municipal supply:	160 gpdpc
Total use—USA:	1800 gpdpc

Permeability	$1\ \mu\text{m}^2 \approx 1$ darcy
Permeability/hydraulic conductivity	$1\ \mu\text{m}^2 \approx 3/4\ \text{m/d} \approx 18.2\ \text{gpd/ft}^2$ at 15.5°C
Darcian validity	$N_R \leq 1$
Specific storage	$S_s \approx 10^{-6}\ \text{ft}^{-1}$ or m^{-1} for sedimentary rocks $S_s \approx 10^{-5}\ \text{ft}^{-1}$ or m^{-1} for unconsolidated sediments
Storage coefficient	$S \approx 10^{-4}$ for sedimentary rocks
Storativity	$S_{\text{conf}} \approx 10^{-4}$ typical $S_{\text{unconf}} \approx 10^{-1}$ typical

Appendix 3

HYDROGEOLOGY EQUATIONS CHEATSHEET

$$n \equiv \frac{V_v}{V_b}$$

$$S_y \equiv \frac{V_w(\text{drained})}{V_b}$$

$$S_r \equiv \frac{V_w(\text{retained})}{V_b}$$

$$n = S_y + S_r$$

$$n = 1 - \frac{\rho_b}{\rho_s}$$

$$S = \frac{V_w}{A\Delta h}$$

$$k = Cd^2$$

$$K = \frac{k\rho g}{\mu} = \frac{k\gamma}{\mu}$$

$$\frac{K_T}{K_{15.5°C}} = \frac{\mu_{15.5°C}}{\mu_T}$$

$$T = Kb$$

$$h = h_p + z$$

$$p = h_p\gamma$$

$$\varphi = hg = h_p g + zg$$

$$Q = KA\frac{dh}{dl}$$

$$V \equiv \frac{Q}{A} = K\frac{dh}{dl}$$

$$v = \frac{V}{n} = \frac{Q}{nA} = \frac{K}{n}\frac{dh}{dl}$$

$$N_R = \frac{\rho V d}{\mu}$$

$$h = h_0 - \frac{Q}{Tw}x$$

$$h^2 = h_0^2 - \frac{2Q}{Kw}x$$

$$\bar{K}_x = \frac{\sum^n K_i z_i}{Z}$$

$$\bar{T}_x = \sum^n T_i$$

$$\bar{K}_z = \frac{Z}{\sum^n \frac{z_i}{K_i}}$$

$$\bar{K} = \sqrt[n]{K_i K_{i+1} \ldots K_n}$$

$$Q = \frac{n_s}{n_d}TH$$

$$\frac{K_1}{K_2} = \frac{\tan\theta_1}{\tan\theta_2}$$

$$\alpha = \frac{dV_t/V_t}{d\sigma_e}(=1/E)$$

$$\beta = \frac{dV_w/V_w}{dP}(=1/E)$$

$$S_s = \gamma(\alpha + n\beta)$$

$$S = S_s b$$

$$\frac{\delta^2 h}{\delta x^2} + \frac{\delta^2 h}{\delta y^2} = \frac{S}{T}\frac{\delta h}{\delta t}$$

$$\frac{\delta^2 h}{\delta r^2} + \frac{1}{r}\frac{\delta h}{\delta r} = \frac{S}{T}\frac{\delta h}{\delta t}$$

$$s = \frac{Q}{4\pi T}\int_u^\infty \frac{e^{-u}}{u}du$$

$$u = \frac{r^2 S}{4Tt}$$

$$W(u) \equiv \int_u^\infty \frac{e^{-u}}{u}du$$

$$s = \frac{Q}{4\pi T}W(u)$$

$$T = \frac{Q}{4\pi s}W(u)$$

$$S = \frac{4\,Tut}{r^2}$$

$$s = \frac{2.3Q}{4\pi T}\log\frac{2.25Tt}{r^2 S}$$

$$T = \frac{2.3Q}{4\pi\Delta s} \qquad S = 2.25T\left(\frac{t}{r^2}\right)_0$$

$$T = \frac{2.3Q}{4\pi\Delta s} \qquad S = \frac{2.25Tt_0}{r^2}$$

$$T = \frac{-2.3Q}{2\pi\Delta s} \qquad S = \frac{2.25Tt}{r_0^2}$$

$$s = \frac{Q}{4\pi T}W(u_A, u_B, \beta)$$

$$T = \frac{Q}{4\pi s}W(u_B, \beta)$$

$$S_y = \frac{4Tt}{r^2}u_B$$

$$K_v = \frac{K_h b^2 \beta}{r^2}$$

$$Q = \frac{2\pi T(h_2 - h_1)}{2.3\log\frac{r_2}{r_1}}$$

$$Q = \frac{\pi K(h_2^2 - h_1^2)}{2.3\log\frac{r_2}{r_1}}$$

Appendix 4

SUMMARY OF EQUATIONS FOR AQUIFER TESTS

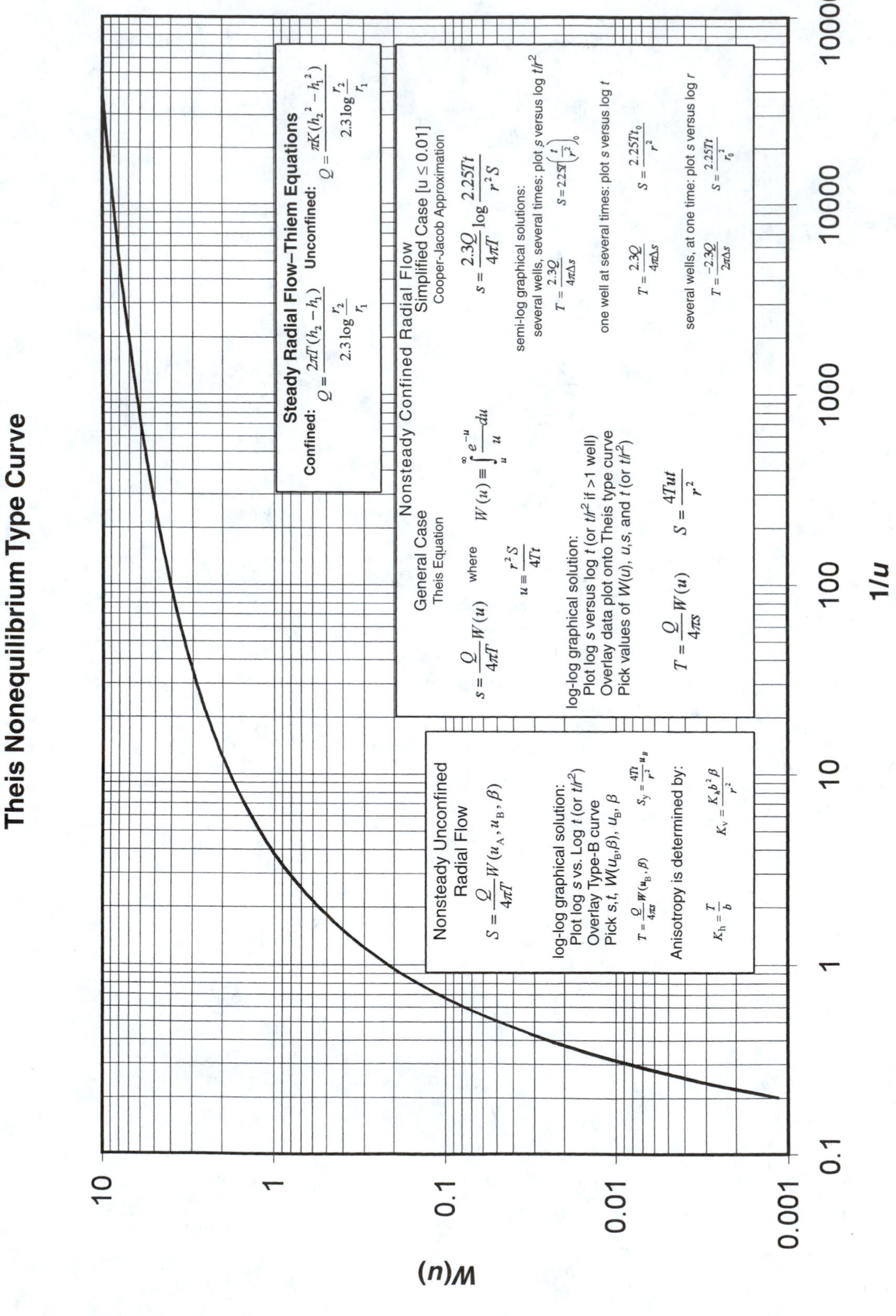